T0358545

Recent Advances in Decolorization and Degradation of Dyes in Textile Effluent by Biological Approaches

Recent Advances in Decolorization and Degradation of Dyes in Textile Effluent by Biological Approaches

Ram Lakhan Singh, Pradeep Kumar Singh and Rajat Pratap Singh

CRC Press
Taylor & Francis Group
Boca Raton London New York

CRC Press is an imprint of the
Taylor & Francis Group, an **informa** business

CRC Press
Taylor & Francis Group
6000 Broken Sound Parkway NW, Suite 300
Boca Raton, FL 33487-2742

© 2020 by Taylor & Francis Group, LLC
CRC Press is an imprint of Taylor & Francis Group, an Informa business

No claim to original U.S. Government works

International Standard Book Number-13: 978-0-367-19952-4 (Hardback)

Visit the Taylor & Francis Web site at
http://www.taylorandfrancis.com

and the CRC Press Web site at
http://www.crcpress.com

Contents

List of Figures

List of Tables

List of Tables

Preface

S YNTHETIC DYES ARE WIDELY used in textile industries. These industries generate a large volume of colored effluent which contains different types of pollutants including dyes. The release of dye-containing effluent into the environment has adverse and toxic effects on all forms of life. Therefore, removal of toxic dyes from textile effluent is necessary prior to their disposal into the environment. Different physical and chemical methods are utilized for removal of dyes from textile effluent. But these methods have several disadvantages which restrict their usage for effective treatment of textile effluent. Biological treatment methods including microorganisms and plants are efficient and eco-friendly in nature and offer several advantages over conventional physical and chemical methods. Nowadays, various advanced methods such as genetic engineering, nanotechnology, immobilized cells or enzyme, biofilms and microbial fuel cells (MFCs) are emerging as potential tools for the treatment of textile effluent.

It gives us immense pleasure to introduce this Focus titled book *Recent Advances in Decolorization and Degradation of Dyes in Textile Effluent by Biological Approaches*. The purpose of this book is to provide updated information regarding different processes used for removal of dyes from textile effluent. This book covers all the details and aspects of textile effluent treatment. The book has been written in a simple and lucid style and is easy to follow with simple explanations and a good framework for understanding various approaches used for treatment of textile effluent. Starting with the basic knowledge of dyes and textile effluent, the book provides brief information on the toxicity of dyes and detailed discussion on various conventional and advanced methods for the treatment of textile effluent. This book will serve as a valuable resource for graduate and undergraduate students, professionals and others who are concerned with the treatment of industrial effluent.

Ram Lakhan Singh
Pradeep Kumar Singh
Rajat Pratap Singh
Ayodhya, Uttar Pradesh, India

Authors

Professor Ram Lakhan Singh currently holds the position of Professor of Biochemistry and Coordinator of the Biotechnology Programme at Dr. Rammanohar Lohia Avadh University, Ayodhya, India. In 1987 he was awarded a PhD by Kanpur University for his extensive work on toxicity of synthetic food dyes and their metabolites at Indian Institute of Toxicology Research, Lucknow, India. Professor Singh joined G.B. Pant University of Agriculture & Technology, Pantnagar, India in 1988 as Assistant Professor of Biochemistry where he taught biochemistry courses to students of Biochemistry, Biotechnology, Microbiology, Agriculture, Home Science and Fisheries. On the research front, he studied the toxicity of pulp and paper mill effluents on plant and animal systems, and also established a toxicology laboratory at the university. He joined Dr. Rammanohar Lohia Avadh University, Ayodhya as Associate Professor of Biochemistry in 1994 and became full Professor in 2002. He developed the undergraduate and postgraduate courses in Biochemistry, Environmental Sciences and Biotechnology.

Professor Singh has guided 25 PhD students for their dissertation work. His main areas of research are Nutraceutical Biochemistry, Environmental Biotechnology and Toxicology. He has published 84 research papers in national and international journals, attended various scientific conferences and chaired scientific/technical sessions. Professor Singh has edited four books, *New and Future Developments in Microbial Biotechnology and Bioengieering: From Cellulose to Cellulase: Strategies to Improve*

Biofuel Production (Elsevier, 2019); *Advances in Biological Treatment of Industrial Waste Water and Their Recycling for Sustainable Future* (Springer Nature, 2019); *Biotechnology for Sustainable Agriculture: Emerging Approaches and Strategies* (Elsevier, 2018); *Principles and Applications of Environmental Biotechnology for a Sustainable Future* (Springer Nature, 2017) and contributed 14 chapters in various books published by international publishers. He is on the panel of experts of academic bodies and selection committees of various universities and funding agencies. Professor Singh has delivered a number of invited/expert talks and popular lectures related to environmental biotechnology/nutraceuticals/toxicology issues on various national and international fora. He is office bearer and life member of several learned societies including: Society of Toxicology (India), Society of Biological Chemists, Indian Science Congress Association, Indian Council of Chemists and Association of Food Scientists and Technologists (India). He was honored as Best Teacher by the International Association of Lions Clubs in 1999. Professor Singh was awarded the International Union of Toxicology (IUTOX) Senior Fellowship by the International Union of Toxicology during the XI International Congress of Toxicology at Montreal, Canada in 2007. He was conferred the Shikshak Shree Samman by the Government of Uttar Pradesh in 2012. Professor Singh was admitted to the Fellowships of the Society of Toxicology, India in 2011 and the Academy of Environmental Biology, India in 2015.

Mr. Pradeep Kumar Singh has been working as Guest Lecturer in the Biotechnology Programme at Dr. Rammanohar Lohia Avadh University, Ayodhya since 2011. He completed his Master's degree in biotechnology from Chhatrapati Shahu Ji Maharaj University, Kanpur in 2009, and worked as Senior Research Fellow in the Uttar Pradesh Council of Agricultural Research (UPCAR) sponsored project entitled "Development of Early Ripening Varieties and Hybrid Varieties of Mustard, Sunflower, Maize and Millet in U.P." at Chandra Shekhar Azad University of Agriculture and Technology

Kanpur, India, from December 2010 to March 2011. He joined the Department of Biochemistry at Dr. Rammanohar Lohia Avadh University, Ayodhya in 2015 for his PhD and worked on bacterial decolorization of textile dyes. Mr. Singh has passed several national exams such as the Council of Scientific & Industrial Research (CSIR) University Grants Commission (UGC National Eligibility Test (NET) in Life Sciences, the Agricultural Scientist Recruitment Board (ASRB) Indian Council of Agricultural Research (ICAR) NET in Basic Plant Science and Agricultural Biotechnology and the Graduate Aptitude Test in Engineering (GATE) in Life Sciences. He has published five research papers in peer-reviewed journals and six book chapters published by international publishers such as Springer Nature and Elsevier. He has participated in various national and international scientific conferences and symposia and presented six research papers. He has guided 14 postgraduate students for their dissertation work.

Dr. Rajat Pratap Singh has been working as Guest Lecturer in the Biotechnology Programme at Dr. Rammanohar Lohia Avadh University, Ayodhya since 2008. He completed his Master's degree in Biochemistry at Dr. Rammanohar Lohia Avadh University, Ayodhya in 2005 and worked as Junior Research Fellow in a Department of Biotechnology (DBT) sponsored project entitled "Development of Sustainable Management Strategies for the Control of *Parthenium* Weeds Using Biotechnological Approaches in U.P." at ICAR – National Bureau of Agriculturally Important Microorganisms, Mau, India from 2005 to 2008. He joined the Department of Biochemistry at Dr. Rammanohar Lohia Avadh University, Ayodhya in 2010 and worked extensively on microbial decolorization of textile effluent and was awarded a PhD degree in 2015. Dr. Singh has passed several national exams such as the CSIR UGC NET, the ASRB ICAR NET and GATE. Dr. Singh edited *Advances in Biological Treatment of Industrial Waste Water and Their Recycling for a Sustainable Future* (Springer Nature Publishing, 2019). He has published six research papers in peer reviewed journals and seven book chapters published by international publishers such as Springer Nature and Elsevier. He has participated in various national and international scientific conferences and symposia, and presented 16 research papers. Dr. Singh has also participated in three national training programmes related to microbial diversity analysis and identification of agriculturally important microorganisms. He has guided 14 postgraduate students for their dissertation work.

CHAPTER 1

Introduction

THE POPULATION EXPLOSION COMBINED with the industrial revolution has led to various environmental concerns. The rapid growth of industries has prompted the production of heavy loads of liquid waste. Environmental pollution, as a result of urbanization and industrialization, has been identified as a noteworthy issue throughout the world which has a detrimental effect on human wellbeing and the biodiversity of the Earth. Food, clothing and shelter are the three fundamental needs of mankind. With the increased demand for textile products, the textile industry, and its effluents have been increasing, making them one of the principle sources of serious pollution around the world. Nature has an impressive ability of coping with small amounts of wastewater and pollution, however, it would be unsafe if the large amounts of effluent produced daily were not treated before discharging them back into the environment. The amounts and characteristics of released effluents vary from industry to industry depending on the water consumption and the average daily product. Textile industries consume large volumes of potable water in various processing stages and produce vast volumes of liquid effluents containing dye

(Andleeb et al. 2010; Singh et al. 2019). About 40–65 liters of effluent is released per kilogram of fabric during the coloring process (Manu and Chaudhari 2002). Textile effluent has high biochemical oxygen demand (BOD), chemical oxygen demand (COD), total dissolved solids (TDS), total suspended solids (TSS), total solids (TS) and also contains degradable and non-biodegradable chemicals such as dyes, dispersants and detergent, surfactants, leveling agents, heavy metal ions, dissolved inorganics, acids and alkali that are discharged into water bodies. Major pollutants and potentially hazardous compounds are discharged from the wet processing stages of textile industries.

Color is the primary contaminant to be seen in textile effluent which is mainly because of the presence of azo dyes. Azo dyes are synthetic aromatics compounds substituted with azo groups (-N=N-). In addition to azo dyes, other classes of synthetic dyes such as anthraquinone, triphenylmethane, phthalocyanine dyes, etc. are also used in various industries. More than 100,000 dyestuffs are available commercially and annual production of the dyes is more than 7×10^5 tons worldwide (Selvam et al. 2003; Singh et al. 2019). These dyes are widely used in textile, paper, leather, food, cosmetic and other industries (Singh et al. 2017a). Due to inefficiency in the dyeing and finishing processes, 10–15 percent of the total dyestuffs are discharged with the effluent into the environment (Khehra et al. 2006). These effluents can alter the physical, chemical and biological nature of the receiving water bodies such as rivers and lakes. This increase in the BOD, COD, TDS and total suspended solids (TSS), alters the pH and gives the intense colorations to water bodies. The colloidal matter present along with the colors increases the turbidity and gives the water a dirty appearance and foul smell. The release of colored textile effluent poses a serious threat to the environment. Discharge of these effluents into water streams not only creates an aesthetic problem but also these are toxic to aquatic life and eventually affect the health of both humans and animals. These colored effluents influence the transparency of water bodies and decrease the rate of photosynthesis by aquatic plants leading to damage of the aquatic environment. Moreover, some azo dyes and their degradation products are a serious threat to the environment due to their toxicity including mutagenicity and carcinogenicity (Wang et al. 2013). The stability and xenobiotic nature of azo dyes makes them recalcitrant. Hence it is necessary to treat the dyes present in textile effluent prior to their disposal. The treatment of polluting dyes is an important issue for small scale textile industries. These small scale industries dump their effluent into the main stream of water resources in light of the fact that their working conditions and economic status do not enable them to treat their wastewater before disposal. Without sufficient and proper treatment these dyes will remain in the environment for an extended period of time.

Various physico-chemical methods are utilized to treat the dye containing textile effluent. Physical methods involve the utilization of adsorbents, coagulants and filtration techniques whereas chemical methods include Fenton oxidation, ozonation, electrochemical oxidation and irradiation. Some of these methods are observed to be effective but they are costly and have other drawbacks (Singh et al. 2017a). These methods are economically unattractive and produce secondary pollution problems due to the generation of a large amount of sludge (Singh et al. 2019). Moreover, these methods are not suitable for degrading the wide variety of dyes. Biological methods are effective, economic, specific, less energy intensive and an eco-friendly alternative for the treatment of textile effluent. The biological treatment processes are based on the degradative abilities of different trophic groups of microorganisms. These microorganisms are capable of decolorizing and degrading the dyes present in the effluents. The biological methods include the use of unicellular microorganisms (bacteria, fungi, algae) and plants (phytoremediation) to remove the dyes from textile effluent (Kirby et al. 2000). The bioremediation of dyes by biological methods is attributed to either biosorption or the enzyme mediated degradation process. Various enzymes which are produced by microorganisms and plants are utilized as a molecular tool for bioremediation of textile dyes with great success. However, although these methods are cost effective and produce less sludge they have some disadvantages. Recently several advanced methods/approaches such as genetic engineering, nanotechnology, immobilized cells or enzymes, biofilms and MFCs, etc. have been utilized to degrade the textile dyes present in effluent and they are genuinely effective. The aim of all of these treatment methods is to set the safety framework for all living organisms and conserve the biodiversity.

1.1 PHYSICO-CHEMICAL CHARACTERISTICS OF TEXTILE EFFLUENT

On the basis of a large volume of wastewater and composition of effluent, textile industries are considered to be one of the most polluting industries. The textile effluent contains organic and inorganic chemicals which are responsible for pollution problems. The major pollutants in textile wastewater originate from dyeing and finishing procedures. Different classes of dyes are utilized for coloring and shading the textiles during the dyeing stage. Dyeing is the procedure of adding color to the textile products such as fibers, yarns and fabrics. Subsequent to coloring, the textiles are subjected to an assortment of finishing processes. These production processes consume vast volumes of water as well as generate a huge amount of wastewater and a generous quantity of waste products (Babu et al. 2007).

These chemicals contribute to the organic load of textile effluent. There are striking changes in various parameters of textile effluent. It is mainly characterized in terms of BOD, COD, dissolved solids, suspended solids, pH, temperature, color, acidity, alkalinity, heavy metals and other soluble substances. Textile effluent produced from dyeing procedures is rich in color and has a very high BOD, COD, TSS and TDS. The TSS is mainly due to organic fractions such as fiber scrap and undissolved raw materials while the addition of inorganic salts contributes to TDS.

1.2 CLASSES OF DYES USED IN THE TEXTILE INDUSTRY

Synthetic dyes are generally water soluble or water dispersible colored organic compounds which are widely used for imparting color to various substrates such as textiles. These dyes are capable of absorbing the light in the visible region (400–700 nm). The chromophore group of dye molecules is responsible for the absorption of light. Dyes can be obtained from natural sources or manufactured artificially. The first synthetic dye, Mauveine, was synthesized by Perkin which was made up from coal tar. Dyes may be classified in a number of ways, according to their chemical structure, color or industrial application (Table 1.1). The classification of all commercial textile dyes has been compiled in the Color Index (CI) which is published by the Society of Dyers and Colourists (United Kingdom) in association with the American Association of Textile Chemists and Colorists (AATCC) (Clark 2011). Table 1.2 shows the chromophore group and structural formula of some textile dyes.

1.3 TOXICITY OF TEXTILE DYES

The wastewater discharge from textile industries contains various hazardous toxic inorganic and organic pollutants including dyes. Because of the way the dyes are synthesized to be chemically and photolytically stable, they are persistent in natural environments (Rieger et al. 2002). Therefore, release of hazardous dyes in the environment can be an ecotoxic risk (Van der Zee 2002). The presence of small amounts of dyes in water is highly visible and reduces photosynthetic activity by lowering the light penetration through water, causing oxygen deficiency and deregulating the biological cycles of aquatic biota as well as limiting downstream beneficial uses, for example, recreation, drinking water and irrigation (Apostol et al. 2012).

Textile dyes can produce an adverse health impact which may be either acute or chronic or both. Many azo dyes can cause acute to chronic effects on organisms, depending on exposure time and

TABLE 1.1   Classification of dyes

Class	Chromophoric types (chemical class)	Fiber types	Examples
Water soluble dyes			
Acid dyes	Nitro and nitroso, triphenylmethane, xanthene, anthraquinone, azine, azo	Silk, wool, nylon, acrylics, also used to some extent for leather, food, paper, cosmetics and ink-jet printing	CI Acid Blue 45, CI Acid Yellow 23, CI Acid Red 73
Basic dyes	Anthraquinone, azo, azine, cyanine, hemicyanine, diazahemicyanine, acridine, diphenylmethane, triarylmethane, oxazine, xanthenes	Modified nylons, polyacrylonitrile, modified polyesters, acrylic, paper, wool, cotton, silk	CI Basic Brown 1, CI Basic Yellow 2, CI Basic Blue 9
Direct dyes	Dioxazine, stilbene, azo, phthalocyanine and other smaller chemical classes such as anthraquinone, formazan, thiazole quinoline	Cotton and other cellulosic fibers, paper, leather and nylon fiber (lesser extent)	CI Direct Orange 26, CI Direct Red 28 (Congo Red), CI Violet 22
Reactive dyes	Phthalocyanine, oxazine, azo, formazan, anthraquinone	Cotton, natural and man made cellulosic fibers, natural protein fibers, silk, nylon, wool, polyamide fibers	CI Reactive Red 31, Procion H, Cibacron E
Water insoluble dyes			
Disperse dyes	Anthraquinone, nitro, benzodifuranone styryl, azo	Acetate and triacetate fibers, nylon, polyester, acrylic	CI Disperse Yellow 3, CI Disperse Red 4, CI Disperse Blue 27
Solvent dyes	Triphenylmethane, anthraquinone, phthalocyanine, azo	Synthetics, plastics, waxes, oils, gasoline	CI Solvent Red 24, CI Solvent Yellow 124, CI Solvent Blue 35

TABLE 1.1 (Cont.)

Class	Chromophoric types (chemical class)	Fiber types	Examples
Sulfur dyes	Indeterminate structures	Man made cellulosic fibers, cotton and rayon	CI Yellow 2, CI Red 5, CI Blue 15
Vat dyes	Indigoid and thioindigoid, anthraquinone	Cotton, polyester, nylon, wool rayon and synthetics wool	CI Blue 4 (Indanthrene), CI Vat Green 9, Tyrian Purple
In-situ color formation			
Azoic dyes	Azo	Cotton, nylon (special purpose), polyester	CI Orange II (Acid Orange 7), CI Acid Red 13, Metanil Yellow

concentration of azo dye. The absorption of azo dyes and their breakdown products such as toxic aromatic amines through the gastrointestinal tract, skin and lungs can potentially cause serious health issues. The skin sensitization effects of azo dyes are common in the textile industry, resulting in occupational eczema. Some azo dyes and their breakdown products are highly mutagenic and carcinogenic (Esancy et al. 1990). Exposure to azo dyes can cause bladder cancer, hepatocarcinomas and nuclear anomalies like chromosomal aberration in mammalian cells (Lima et al. 2007). The azo dyes are chiefly metabolized in the intestine and liver, resulting in the production of aromatic amines that are potentially carcinogenic and mutagenic (Gingell et al. 1971). In mammals, metabolic activation (reduction of azo linkage) of azo dyes is mainly because of bacterial action in the anaerobic parts of the lower gastrointestinal tract (Van der Zee 2002). Azo dyes can cause damage to DNA which prompts development of malignant tumors. The cancer causing capability of the dye increased when electron donating substituents are present in ortho and para positions. In mortality tests with rodents, the majority of azo dyes showed LD_{50} values between 250 and 2000 mg/kg body weight (Van der Zee 2002). Consequently, the chances of human mortality

TABLE 1.2   Some textile dyes, their chromophore group and structure

Textile dye	Chromophore group	Structure
Acid Red 138	Azo	
Alizarin	Anthraquinone	
Malachite Green	Triphenyl-methane	
Naphthol Green-B (Acid Green 1)	Nitro	
Pigment Green 7	Phthalocyanine	
Tyrian Purple	Indigo	

due to acute dyestuff exposure are probably low. Algal growth and fish mortality are not influenced by dye concentrations below 1 mg/L. The most toxic dyes for algae and fishes are acid and basic dyes with a triphenylmethane structure (Greene and Baughman 1996). Azo dye inclination to bioaccumulation has been broadly investigated in fish. The absorption rates of dyes are influenced by their partition coefficient (Erickson and McKim 1990). Water soluble dyes like acid, reactive and basic dyes are generally not bioaccumulated.

Some reactive dyes are perceived as respiratory sensitizers. These reactive dyes can cause respiratory diseases and allergic dermatoses (Nilsson et al. 2006). Breathing in respiratory sensitizers can cause occupational asthma. Some dyes can cause unfavorably susceptible skin responses. The workers exposed to reactive dyes have an increased risk of colon and rectum cancer as well as bladder cancer (De Roos et al. 2005). Certain reactive, vat and disperse dyes are perceived skin sensitizers. Anthroquinone dyes have a fused aromatic ring structure which is resistant to degradation. Also, anthroquinone dye such as Malachite Green has an adverse impact on immune and reproductive systems. It is also a genotoxic and carcinogenic agent and may cause fragmentation of DNA.

In addition to textile dyes, different textile chemicals such as ammonia, acetic acid, soda ash, caustic soda and bleaching agents used in various processing stages of textiles can also cause health problems. These chemicals may cause skin irritation, itchy or blocked noses, sneezing and sore eyes. The textile effluent also contains some heavy metals in trace concentration such as chromium, lead, nickel, copper and zinc which are responsible for causing several health issues including hemorrhaging.

The toxicity of dyestuff has been evaluated using various methodologies with aquatic organisms (fish, algae, bacteria, etc.) in mammals and plants (Singh et al. 1987, 1988, 1991; Singh 1989; Jadhav et al. 2011) (Figure 1.1).

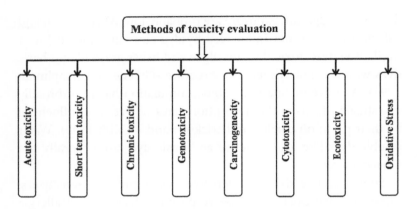

FIGURE 1.1   Methods to evaluate toxicity of textile dyes and their breakdown products

CHAPTER 2

Methods of Decolorization and Degradation of Dyes in Textile Effluent

FOR SEVERAL YEARS MANY physical and chemical methods (Table 2.1) have been used to treat the dyes containing textile effluent. Some of them are very effective and have been used at a commercial level to remove the harmful dyes from textile effluent. In order to provide new treatment alternatives that are superior to physico-chemical methods, the use of living organisms (biological methods) are recommended.

2.1 PHYSICO-CHEMICAL METHODS – AN OVERVIEW

Among physical methods the adsorption process is promising for treating textile effluent. Various types of adsorbents, including synthetic and natural ones such as laumontite, ferrierite, alumina,

TABLE 2.1   Advantages and disadvantages of physico-chemical methods for decolorization and degradation of textile effluent

Physical/chemical method	Advantages	Disadvantages
Coagulation/ Flocculation	Low cost	Large amount of sludge produced
Cucurbituril	Adsorb several types of dyes	High cost
Fenton's reagent	Removes both soluble and insoluble dyes	Excess sludge generated
Ion exchange	No loss of ion exchangers	Not effective against all classes of dyes
Membrane filtration	Effective against all classes of dyes	Concentrated sludge produced
NaOCl	Breaks azo bond	Produces toxic aromatic amines
Ozonation	Applied in gaseous state	Half life is very short, releases carcinogenic nitrogenous aromatics
Photochemical	No sludge generated	Ends with by-product formation
Silica gel	Suitable for removal of basic dyes	Commercial application is not feasible
Wood chip	Suitable for adsorption of acid dyes	More time consuming process

activated carbon and silica, are commonly used in the adsorption process. In this process textile dye is transferred at the interface between two immiscible phases in contact with one another. The ion exchange technique is an efficient method of removing anionic and cationic dyes. Extensive use of this method is not advised for textile wastewater treatment because ion exchangers cannot hold a wide range of dyes. Synthetic porous crystalline gel alumina and porous, non crystalline silica gel of different sizes prepared by the coagulation of colloidal silicic acid have been used for successfully to remove dyes (Huang et al. 2007). The irradiation process is used to treat textile effluent in a dual tube bubbling reactor in the presence of a large volume of dissolved oxygen (Kumar et al.

1998). Membrane filtration is a physical technique that is widely used for wastewater treatment. The filtration process includes microfiltration, ultrafiltration, nanofiltration and reverses osmosis. Ultrafiltration, nanofiltration and reverse osmosis are used for removal of dyes from dye bath wastewater and bulk textile processing wastewater (Kim and Shoda 1999).

Several chemical methods are also widely used for removal of dyes from textile effluent. Usually oxidizing agents are used to facilitate chemical oxidation (Ranganathan et al. 2007). The electrochemical method is a newer, highly advanced technique to remove color and degradation of recalcitrant pollutants from dye wastewater (Pelegrini et al. 1999). This process is based on the supply of an electric current to wastewater by using sacrificial iron electrodes to produce ferrous hydroxide to remove soluble and insoluble acid dyes from the wastewater. The process of ozonation, ozone mediated oxidation, is used to degrade phenols, pesticides, aromatic hydrocarbons and chlorinated hydrocarbons (Xu and Leburn 1999). Ozone quickly decolorizes water soluble dyes but the removal of non soluble dyes takes longer. Fenton's reagent (solution of hydrogen peroxide (H_2O_2) and an iron (Fe) catalyst) is an appropriate reagent for the treatment of textile wastewater. This oxidizing reagent is mainly useful for the treatment of effluents that are resistant to biological treatment and poisonous to live biomass. Fenton's reagent can decolorize a variety of textile dyes; it is cheap and is more effective at removing COD (Ince and Tezcanli 1999). The photochemical method catalyzes the degradation of organic dye molecules to carbon di oxide (CO_2) and water (H_2O) by ultraviolet (UV) treatment in the presence of H_2O_2 (source of hydroxyl radicals upon activation). UV light is used to activate H_2O_2. The rate of dye removal largely depends on pH, dye structure and the intensity of the UV radiation used. The UV light has also been tested in combination with Fenton's reagent, titanium dioxide (TiO_2) and ozone (O_3) and other available solid catalysts for the decolorization and degradation of dye water (Hao et al. 2000). In developed nations, the coagulation-flocculation is the most commonly used method for treatment of textile wastewater. The major advantage of this method is that it can be either used as a pre-treatment and post-treatment or as a main treatment system (Gahr et al. 1994). Organic and inorganic (alum, lime and iron salts) polymer coagulants are used either independently or in a mixture with one another to treat dye wastewater for color removal (Chen et al. 2012). This method can competently decolorize sulfur particularly and disperse dyes, however acid, direct, reactive and vat dyes are not easily removed by coagulation-flocculation (Vandevivere et al. 1998).

2.2 BIOLOGICAL METHODS

Various types of pollutants are discharged by textile industries which pose a serious threat to environmental safety. The main pollutants of textile effluent are azo dyes which are recalcitrant in nature (Singh et al. 2015a). The existing effluent treatment processes are not able to completely remove the recalcitrant azo dyes from textile effluent. In recent years, processes based on bioremediation have been seriously considered for the treatment of textile dyes. Biological treatment methods are often the preferred choice because of their low environmental impact and costs in comparison with other treatment technologies (Carneiro et al. 2010). The biological treatment methods depend on the ability of microorganisms and plants to remove the textile dyes from textile effluent. In the environment, there are a large number of microorganisms the potential of which is not utilized completely. Microorganisms can break down most of the compounds in textile effluent and use them for their growth and/or energy needs. In some cases, the metabolic pathways of microorganisms may also be used to break down the pollutant molecules in order to assist the normal growth and development of the microorganisms. In biological treatment, microorganisms acclimatize themselves to toxic wastes such as dyes and transform them into nontoxic or less harmful forms (Saratale et al. 2009). Microorganisms that have the ability to degrade azo dyes are necessary for the efficient treatment of textile effluent. A wide range of microorganisms are capable of degrading azo dyes that belong to different taxonomic groups of bacteria, fungi and algae under certain environmental conditions (Khehra et al. 2005). The versatility of microorganisms makes them capable of degrading a wide variety of dyes. These methods are based on the microbial biotransformation of dyes. Microbial systems have the ability to degrade the textile dyes as well as other pollutants which ultimately decreases the BOD and COD of textile effluent. The effectiveness of microbial decolorization of azo dyes depends on the adaptability and the activity of the selected microorganisms, dye class, dye concentration and physic-chemical characteristics of textile effluent such as pH, temperature and the presence of organic and inorganic pollutants.

2.2.1 Mechanism of Decolorization and Degradation of Dyes in Textile Effluent

The microbial decolorization and degradation of dyes in textile effluent can be achieved by using one of two mechanisms: biosorption and enzymatic degradation.

2.2.1.1 Biosorption

The uptake or accumulation of chemicals such as dyes by microbial biomass or biological materials is known as biosorption (Singh and Singh 2017). The biosorption process involves a solid phase (biosorbent) and a liquid phase which contain a dissolved species to be sorbed such as dyes. The ions or molecules present in one phase (the liquid phase) are slanted to accumulate and concentrate on the surface of another phase (the solid phase). Adsorption of the dye onto the biomass is the simplest mechanism of color removal. A wide variety of microorganisms belonging to various groups (bacteria, fungi, yeasts and plants) have been reported to remove a broad range of dyes by biosorption. Biosorption, with the help of microbial systems, involves the binding of pollutants (dye) to the cell surface through physical adsorption, ionic and chemical interactions, ionic exchange and chelation. The microbial cell wall contains different functional groups such as hydroxyl, carboxyl, amino and phosphate groups which involve an interaction between the cell wall and the dye. These functional groups specify the biosorption capacity of a microorganism (Charumathi and Das 2010). The biosorption process changes the phase of pollutants from one phase to another and thus produces sludge which needs to be disposed of safely. The biosorption process is not appropriate for long-term treatment in light of the fact that during adsorption, the dye is concentrated onto the biomass, which will become saturated over time. Dead microbial cells have advantages as biosorbents over living cells as they do not require nutrients, can be stored and utilized for longer periods and can be recovered utilizing organic solvents or surfactants (Fu and Viraraghavan 2001). The efficiency of the biosorption procedure is influenced by pH, temperature, ionic strength, time of contact, adsorbent and dye concentration, dye structure and type of microorganisms (Fu and Viraraghavan 2001).

2.2.1.2 Enzymatic

Azo dyes are electron-deficient compounds because of their azo bridge (-N=N-). The majority of the azo dyes have electron withdrawing groups, which generate an electron deficiency and make them resistant to degradation by the microbial system (Kuberan et al. 2011). Under suitable conditions, these dyes can be degraded by microbial enzymes (Sugumar and Thangam 2012).

2.2.1.2.1 Enzymes Involved in Decolorization and Degradation of Azo Dyes   The azo dye degrading microbes have developed enzyme systems that mediate the degradation of the azo dye by cleaving the azo bond. These microorganisms produce different types of reductive and oxidative enzymes. Usually, the oxido-reductive enzymes generate free radicals that undergo a complex series of cleavage reactions resulting in the decolorization and degradation of azo dyes (Saratale et al. 2009). Among the range of reductive and oxidative enzymes produced by microbes, azoreductases, laccases and peroxidases are the main enzymes for decolorization and degradation of azo dyes (Chacko and Subramaniam 2011).

2.2.1.2.2 Reductive Enzymes

2.2.1.2.2.1 Azoreductases Azoreductases are considered to be the main enzyme in decolorization and degradation of azo dyes. These enzymes catalyze the reductive cleavage of azo bridge to their corresponding aromatic amines with the help of reducing equivalents such as NADH, NADPH and $FADH_2$ (Chacko and Subramaniam 2011; Singh et al. 2017b). These aromatic amines are generally colorless, hence the reduction of azo dye is also known as decolorization. Different types of azoreductases have been isolated and characterized from different organisms for removal of azo dyes (Ramalho et al. 2005). These enzymes are either specific to a particular dye or catalyze the reduction of a wide range of dyes. Azoreductases are flavoproteins in structure and are diverse in nature and function. Because of low homology in primary amino acid level, it is hard to classify the azoreductases on the basis of their primary amino acid level. Nonetheless, Abraham and John (2007) have classified the azoreductases on the basis of their secondary and tertiary amino acid analysis. On the basis of function, azoreductases are classified as flavin-dependent azoreductases or flavin-independent azoreductases (Blumel and Stolz 2003; Chen et al. 2005; Singh et al. 2015b). The flavin-dependent azoreductases utilize the reducing equivalents (NADH, NADPH or both) as electron donor (Chen et al. 2004; Chen et al. 2005; Wang et al. 2007). The cleavage of azo linkage by azoreductases is a four electron four proton reduction reaction which is carried out in two sequential steps. In the first step, two electrons are transferred to the azo dye with the help of reducing equivalents forming a hydrazo intermediate (Singh et al. 2019). This hydrazo intermediate further breaks into aromatic amines in the second step by the transfer of the next two electrons (Deller et al. 2006).

The mechanism by which azoreductase enzymes transfer electrons to the dye molecules may involve direct or indirect enzymatic reduction. The direct bacterial enzymatic reduction mechanism includes the direct physical interaction of azoreductase with azo dye in order to transfer electrons from reducing equivalents which is generated by the oxidation of the substrate/coenzyme (Figure 2.1) (Bragger et al. 1997). The indirect bacterial enzymatic reduction mechanism involves the transfer of electrons by a redox mediator that acts as an electron shuttle between the dye and the azoreductase (Chacko and Subramaniam 2011). The reduced

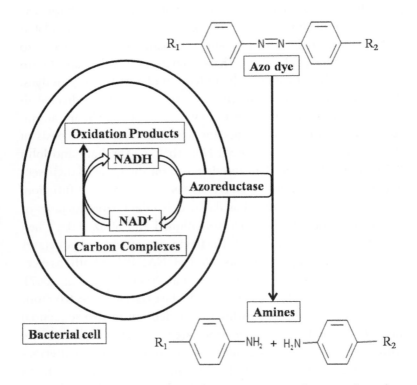

FIGURE 2.1 The mechanism of direct enzymatic reduction of azo dyes by bacteria

electron shuttles can transfer electrons to electron-deficient azo dyes (Watanabe et al. 2009).

The non-sulphonated azo dyes can pass through the cell membrane and their decolorization might be catalyzed by soluble flavin dependent cytoplasmic azoreductases. The majority of the azo dyes are polar and high molecular weight complex compounds which have sulphonate substituent groups. These dyes are unable to diffuse through the cell membranes, therefore reduction of such azo dye is not reliant on their intracellular uptake (Pandey et al. 2007; Singh and Singh 2017; Singh et al. 2019). The bacterial cell membranes are also nearly impermeable to flavin containing cofactors which limit the transfer of reducing equivalents by flavins from the cytosol to the sulphonated azo dyes (Russ et al. 2000). Hence, another mechanism (indirect enzymatic reduction) might be responsible for the decolorization of high molecular weight sulphonated azo dyes where bacteria must set up a link between their intracellular electron transport systems and azo dye in the extracellular environment. The bacterial systems that possess electron transport systems in their membranes can build up the type of link that enables them to interact with the sulphonated azo dyes or a redox mediator (RM) on the cell membrane (see Figure 2.2). Hong et al. (2009) observed that the membrane fraction of *Shewanella decolorationis* S12 contains membrane bound azoreductases and all the other components which facilitate the transfer of electrons from electron donors to the azo compounds with or without the help of redox mediators resulting in the decolorization of azo dye. The redox mediators are low molecular weight compounds which mediate the interaction of membrane bound azoreductases and sulphonated azo dyes. These redox mediators are produced as metabolic products of specific substrates during the bacterial metabolism or they may be added externally (Singh et al. 2015b).

In addition to the azoreductase enzyme, NADH-DCIP reductases and riboflavin reductases are likewise proficient to decolorize and degrade the azo dyes from textile effluent.

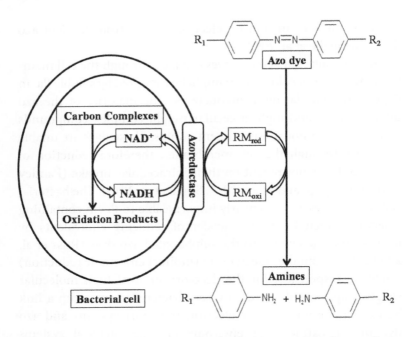

FIGURE 2.2 The mechanism of indirect enzymatic (redox mediator (RM)-dependent) reduction of azo dyes by bacteria

2.2.1.2.3 Oxidative Enzymes   Some oxidative enzymes such as laccase and peroxidase (lignin peroxidase, manganese peroxidase) are also capable of decolorizing and degrading azo dyes.

2.2.1.2.3.1 Laccases   Laccases are copper containing enzymes which belong to the family of copper containing polyphenol oxidases (Singh et al. 2015b). They are also known as multicopper phenol oxidases (Giardina et al. 2010). Laccases are the oxidoreductases found in higher plants, fungi, bacteria and insects. These enzymes exhibit a broad range of substrate specificity and have the capability to degrade a range of xenobiotics with different chemical structures. Laccases have great potential in bioremediation, because of their particular properties such as nonspecific oxidation capacity, lack of necessity for cofactors and utilization of accessible oxygen as an electron acceptor (Telke et al. 2011). Laccases have been studied in some depth on account of their capacity to decolor and degrade azo dyes present in textile effluent. These enzymes are mainly extracellular and decolorize the azo dyes through a highly nonspecific free radical mechanism without direct cleavage of the azo linkage which avoids the production of hazardous intermediates (aromatic amines).

They catalyze the monoelectronic oxidation of different substrates such as phenolic compounds in the presence of oxygen as an electron acceptor (Sharma et al. 2007). The oxidation of phenolic compounds starts with the formation of unstable phenoxy radicals due to the deprotonation of the hydroxyl group of phenols which leads to the formation of quinine or involves depolymerization and demethylation. Laccases are generally unable to oxidize non-phenolic compounds or molecules with high redox potential (Giardina et al. 2010). The presence of low molecular weight compounds (redox mediators) can overcome the oxidation of such compounds. These redox mediators are oxidized by laccase into stable radicals which are able to oxidize the high redox potential substrate (Arora and Sharma 2010). The redox mediators can be natural, for example, 4-hydroxybenzoic acid, 4-hydroxybenzyl alcohol, 3-hydroxyanthranilic acid produced by lignolytic fungi, or synthetic, for example, 2, 2'-azinobis (3-ethylthiazoline-6-sulphonic acid) (ABTS), 1-hydroxybenzotriazol (Khlifi et al. 2010; Moreira et al. 2014). Laccases are capable of degrading azo dyes in the presence or absence of redox mediators (Moya et al. 2010). The redox mediators have a stimulatory impact on the degradation of textile dye and improve the adequacy of laccases (Grassi et al. 2011).

2.2.1.2.3.2 Peroxidases   Peroxidases are heme containing oxidoreductases found in plants, microorganisms and animals (Duarte-Vazquez et al. 2003). They catalyze the chemical reactions in the presence of hydrogen peroxide as a terminal electron acceptor rather than oxygen (Duran et al. 2002). Peroxidases are notable for their role in the degradation of azo dye. Both extra and intra cellular peroxidases are involved in the degradation of textile dyes. These peroxidases are able to catalyze the degradation of dyes without the formation of hazardous aromatic amines. The peroxidases involved in the degradation of textile dyes include lignin peroxidases (LiPs) and manganese peroxidases (MnPs). Both enzymes have been utilized for the degradation of synthetic dyes (Dawkar et al. 2009).

LiPs and MnPs have a similar reaction mechanism during their catalytic cycle which begins with the oxidation of the enzyme by H_2O_2 to an oxidized state. The LiP mediated catalytic mechanism of azo dye degradation involves the oxidation of the phenolic ring by H_2O_2-oxidised LiP in two successive monoelectronic oxidation steps. This oxidation produces a radical cation (carbonium ion) at the carbon which bears the azo bridge. The nucleophilic attack by water at the phenolic carbon leads to the formation of quinone and the corresponding phenyldiazene. The phenyldiazene can further be oxidized by O_2 via a monoelectronic oxidation reaction to a phenyl radical and the azo linkage is eliminated as N_2 (Chivukula et al. 1995). The MnP mediated mechanism of textile dye degradation involves the oxidation of Mn^{2+} to Mn^{3+} by H_2O_2-oxidised MnP. Mn^{3+} is then stabilized by organic acids and forms an Mn^{3+}-organic acid complex. This complex acts as an active oxidant which is able to oxidize the textile dyes.

2.2.2 Use of Microorganisms in Decolorization and Degradation of Dyes in Textile Effluent

Different groups of microorganisms including bacteria, fungi and algae have the potential to decolorize and degrade the textile dyes present in textile effluent under favorable conditions.

2.2.2.1 Bacteria

Bacteria are the most frequently applied microbes for the removal of dyes from textile effluent because they are easy to isolate and adapt. Moreover, they survive in adverse conditions and grow more quickly than other microorganisms (Solis et al. 2012; Singh et al. 2019). The isolation of potent dye degrading bacterial species is one of the fascinating biological aspects of textile effluent treatment. These bacteria can be isolated from soil, effluent, sludge, contaminated water, human and animal faeces and other ecological niches. The bacterial species isolated from textile effluent contaminated sites have the added advantage that they can adjust to the adverse conditions for decolorization of textile dyes. The process of dye decolorization by bacteria may be anaerobic, aerobic or involve a combination of both. However, there are clear differences between the physiology of bacteria grown under aerobic and anaerobic conditions. The application of the bacterial system in the removal of dyes is a simple and attractive strategy because of its inexpensive and eco-friendly nature. Bacteria have the ability to decolorize and mineralize a variety of textile dyes with their evolved enzyme systems under certain environmental conditions (Dhanve et al. 2008). The bacterial culture exhibits specific cultural and nutritional requirements for its growth and decolorization process. The effectiveness of bacterial decolorization of textile dyes is greatly affected by the operational parameters. These operational parameters are aeration, temperature, pH, dye classes and structure, carbon and nitrogen sources as well as the composition of textile effluent. In addition to azo dyes, textile industries also utilize other classes of dyes such as triphenylmethane and anthraquinone dye. The bacterial species are likewise able to decolorize the triphenylmethane and anthraquinone dyes.

Textile dyes are the main pollutant in textile effluent. In spite of the textile dyes, textile effluent has various other pollutants and high BOD and COD. The majority of the studies have been based on microbial decolorization of textile dyes using dye solutions. However, it is crucial to perform the decolorization experiment with real textile effluent that is complex in nature. Bacterial systems are also capable of removing these pollutants along with azo dyes and decreasing the BOD and COD level of effluents.

2.2.2.1.1 Pure Bacterial Culture   The utilization of a pure bacterial culture to decolor textile dyes ensures that the information is reproducible, and a fruitful explanation of experimental perceptions is simplified. It is likewise helpful in understanding the detailed mechanisms and kinetics of dye degradation using the biochemical and molecular tools. These tools can be used to control the catalyst framework in order to produce modified strains with increased enzyme activities (Saratale et al. 2011). Bacterial decolorization of azo dyes begins with the azoreductase mediated reductive cleavage of highly electrophilic azo linkage under anaerobic conditions. This reductive cleavage produces colorless toxic aromatic amines which are generally resistant to further anaerobic degradation. These aromatic amines can be degraded to nontoxic forms under aerobic conditions (Singh et al. 2015a). The bacterial decolorization of azo dyes under anaerobic conditions is a simple and nonspecific process but the complete mineralization of dye molecules is difficult. Bacterial decolorization of textile dyes under anaerobic conditions is dependent on the organic carbon or energy source since they cannot use the dye as the substrate for their growth and energy (Stolz 2001; Khehra et al. 2005; Singh et al. 2019). Various pure bacterial cultures such as *Acinetobacter calcoaceticus, Bacillus cereus, Pseudomonas aeruginosa* and *Staphylococcus hominis* have been accounted for their role in decolorization of textile dyes under anaerobic conditions (Ghodake et al. 2009a; Modi et al. 2010; Sarayu and Sandhya 2012; Singh et al. 2014). Pokharia and Ahluwalia (2016a) observed that the bacterial strain *Staphylococcus epidermidis* MTCC 10623 showed 99.6 percent decolorization of Basic Red 46 dye after six hours of incubation under static conditions. Chen et al. (2003) found that the bacterial isolate *A. hydrophila* was capable of decolorizing the 22 azo dyes in the range of 51–100 percent after seven days of incubation under anaerobic conditions. The bacterium was also capable of decolorizing anthraquinone (Acid Blue 264) and indigoid dye (Acid Blue 74) under anaerobic conditions in the range of 80–84 percent. The ability of pure bacterial culture to treat the real textile effluent was also reported. The *Pseudomonas* sp. SU-EBT showed high potential to remove textile dyes (90 percent) as well as COD (50 percent) and BOD (45 percent) from raw effluent under static conditions (Telke et al. 2010). Dawkar et al. (2009) reported that the bacterial strain *Bacillus* sp. VUS was capable of removing the color from textile effluent to the extent of 80 percent in 12 hours.

Under aerobic conditions, azo dyes are generally resistant to bacterial degradation because the presence of oxygen hinders the reduction activity of azo linkage (Ola et al. 2010). However, few aerobic bacterial strains have the ability to break the azo bond in oxygen-rich conditions with the help of oxygen-insensitive or aerobic azoreductases which leads to the production of aromatic amines (Kodam et al. 2005). These amines can be further metabolized under aerobic conditions with the help of mono- and di-oxygenase enzymes which catalyze the hydroxylation pathway resulting in (Sarayu and Sandhya 2012). Bacterial decolorization of azo dyes under aerobic conditions requires additional organic carbon and energy sources because the bacteria cannot utilize the azo dye as a growth substrate (Stolz 2001). Ito et al. (2018) observed that the *Bacillus amyloliquefaciens* has the capability of decolorizing the anthraquinone dyes under shaking conditions. The bacterial isolate *Aeromonas hydrophila* was found effective for the decolorization of three triarylmethane dyes namely Basic Violet 14, Basic Violet 3 and Acid Blue 90 under shaking conditions. The decolorization extent was observed to be in the range of 72– 96 percent (Ogugbue and Sawidis 2011).

2.2.2.1.2 Mixed Bacterial Culture/Consortia   The application of a mixed microbial population or consortia for the treatment of textile dyes present in textile effluent offers significant advantages over the utilization of an individual bacterial strain. The consortia can achieve a greater degradation and mineralization of textile dyes due to the synergistic or co-metabolic activities of the microbial community (Khehra et al. 2005; Singh et al. 2015a; Singh et al. 2019). In the microbial consortium, the individual strains may attack the dye molecule at various positions or a degradation intermediate produced by a strain might be utilized by a co-existing strain for further degradation (Jadhav et al. 2008). To achieve the complete mineralization of textile dyes, the microbial population of a consortium should work in both anaerobic and aerobic conditions. The use of a bacterial consortium for the decolorization of azo dyes has been reported in several studies. Khehra et al. (2005) reported the complete decolorization of Acid Red 88 by a consortium of HM-4 (*Bacillus cereus, Pseudomonas putida, Pseudomonas fluorescence* and *Stenotrophomonas acidaminiphila*) in 24 hours due to the synergistic metabolic activity of isolates whereas complete decolorization of Acid Red 88 by monocultures was observed in 60 hours. Cheriaa et al. (2012) observed that the decolorization of Malachite Green and Crystal Violet by a consortium of *Agrobacterium radiobacter, Bacillus* sp., *Sphingomonas paucimobilis* and *Aeromonas hydrophila* was 91 percent and 99 percent, respectively under shaking conditions. The mixed consortium of *Pseudomonas desmolyticum* NCIM 2112 and *Galactomyces geotrichum* MTCC 1360 have the capability to decolorize the anthraquinone dye (Vat Red 10) under aerobic conditions (Gurav et al. 2011). Ayed et al. (2011) found that the bioaugmentation of a bacterial consortium consisting of *Sphingomonas paucimobilis, Bacillus* sp. and filamentous bacteria resulted in the removal of dye from textile effluent up to 86 percent with a concomitant COD reduction of up to 75.06 percent.

2.2.2.2 Fungi

Bacteria and fungi are the main degraders of organic matter, but fungi are more frequently used for this function due to their dominancy in enzyme production. Fungi produce dye degrading enzymes such as MnP, LiP and laccase (Bergsten-Torralba et al. 2009). *Phanerochaete chrysosporium,* white rot fungus is the most widely studied organism for decolorization and degradation of textile effluent. White rot fungi are able to remove dyes by using reductive enzymes such as LiPs, MnPs and oxidative enzymes such as laccases. Azo dyes are the largest class of textile dyes which are not readily degraded by microorganisms but they can be removed by *P. chrysosporium* (Paszczynski and Crawford 1995). Several other fungi reported to have potential to decolorize dye containing wastewater are *Inonotus hispidus, Hirschioporus larincinus, Phlebia tremellosa* and *Coriolus versicolor.* Kirby (1999) has studied the capability of *P. chrysosporium* to decolorize artificial textile wastewater and found 99 percent decolorization within seven days. Namdhari et al. (2012) tested the decolorization potential of the three fungal species for Reactive Blue MR dye in carbon limited Czapek Dox broth and after ten days of incubation *Aspergillus allhabadii, Aspergillus sulphureus* and *Aspergillus niger* decolorized the dye up to 95 , 93 and 83 percent, respectively. Selvam and Shanmuga (2012) concluded that basidiomycetes fungi (*Schizophyllum commune* and *Lenzites eximia*) remove textile dyes present in wastewater by the adsorption process. He et al. (2018) concluded that the high and low level of MnP and LiP activities respectively were responsible for the decolorization of azo dyes from the supernatant of the *Trichoderma tomentosum* culture. Furthermore, two fungal isolates *Aspergillus flavus* and *Aspergillus wentii* were studied for their ability to decolorize an aqueous solution of the dyes Acid Blue and Yellow MGR and textile effluent. The maximum decolorization achieved was 98.47 percent in effluent, 94.41 percent by immobilized *A. flavus* in Acid Blue dye and 95.88 percent by immobilized *A. wentii* in Yellow MGR dye (Dorthy et al. 2012). Similarly, Praveen and Bhat (2012) studied decolorization of azo dye Red 3 BN by three fungal species *P. chrysogenum, A. niger* and *Cladosporium* sp. by using potato dextrose agar medium containing 0.01 percent of Red 3 BN. The decolorization extent recorded by *P. chrysogenum* under ideal conditions was 99.56 percent, *A. niger* was 98.64 percent and that by *Cladosporium* sp. was

98.18 percent. Another fungal isolate *Aspergillus terreus* SA3 isolated from the wastewater showed potential for the removal of dye Sulphur Black from textile effluent (Andleeb et al. 2010).

2.2.2.3 Algae

Algae show xenobiotics binding abilities due to the presence of functional groups such as carboxyl, amino, hydroxyl and sulphates on the surface of their cells. Dead microalgae are effective bio-sorbent material since they are easier to cultivate and have a larger surface area. Algae serve as one of the best biomaterials as they have a high capacity for removing dye from textile effluent (Satiroglu et al. 2002). Algae utilize three intrinsically different mechanisms for color removal, i.e. (i) for the production of algal biomass, CO_2 and H_2O (ii) alteration of colored compound to non-colored compounds and (iii) adsorption of chromophores on algal biomass. Many species of *Chlorella* and *Oscillatoria* successfully degrade azo dyes into their corresponding aromatic amines which can further metabolize into simpler organic compounds such as CO_2, NH_3 and H_2O (Acuner and Dilek 2004). Algae produce enzyme azoreductase which is responsible for the biodegradation of azo dyes (Liu and Liu 1992). Ozer et al. (2006) have studied the potential of dried *Spirogyra rhizopus* to decolorize Acid Red 274 dye by both the biosorption and the biocoagulation process. Daneshvar et al. (2007) investigated the potential of green algae, *Cosmarium* sp. as a viable bio-material for the decolorization and degradation of triphenylmethane dye and Malachite Green. Thermophilic blue green algae *Phormidium* in immobilized conditions shows good decolorization activity under thermophilic conditions (Ertugrul et al. 2008). Lee and Chang (2011) tested the capability of marine algae *Isochrysis galbana*, *Chlorella marina*, *Tetarselmis* sp, *Nannochloropsis* sp. and *Dunaliella salina* and fresh water microalgal cells (*Chlorella* sp.) in dye decolorization from the textile wastewater. The highest decolorization was observed by *Isochrysis galbana* (55 percent) followed by freshwater *Chlorella* sp. (43 percent). Azza et al. (2013) demonstrated the biosorption activities of *Spirogyra* sp. for Basic Blue 3, Basic Red 46 and Acid Orange 7. Kumar et al. (2014) studied the biosorption potential for dye Acid Black 1 by using *Stoecospermum marginatum*, *Nizamuddin zanardini* and *Stoepermum glaucescens* and recorded dye removal up to 97.62 percent, 99.27 percent and 98.12 percent, respectively. Masoud et al. (2012) revealed the optimized conditions of two brown macro algae *Stoechospermumm arginatum* and *Stoechospermum glacescens* for removal of dye Acid Black 1. Ayca et al. (2006) studied the biosorption capacity of *Cyanobacterium* and *Nostoc limckia* HA 46 for Reactive Red 198. The highest biosorption capacity of algal biomass was 94 percent at pH 2 and 35°C with an initial dye concentration of 100 mg/l. Rammel et al. (2011) studied the absorption ability of *Chactophora elegans* for Crystal Violet from an aqueous solution. The algal biomass showed maximum adsorption capacity at 25°C. Khataee et al. (2010) studied the biodegradation of Malachite Green in aqueous solution by macroalgae *Chara*. The decolorization was largely dependent on optimized parameters including pH, temperature, reaction time, initial dye concentration and the adsorbent dose. Lim et al. (2010) used the algae *Chlorella vulgaris* for the bioremediation of textile wastewater and found that the algae remove phosphates, nitrates and COD to the extent of 33.3 percent, 45 percent and 62.3 percent, respectively from textile wastewater. Both living and dead algae have been equally useful in color removal from textile wastewater. Jadhav et al. (2010) demonstrated the biosorption potential of the dead biomass of *Spirogyra* for removal of reactive dye from textile wastewater. Mohan et al. (2002) proposed a mechanism of biosorption during the bio-removal of Reactive Yellow 22 by using *Spirogyra* sp. Guolan et al. (2000) demonstrated the ability of four different algal species (*Spirogyra*, *Chlorella vulgaris*, *Chlorella pyrenoidosa* and *Oscillatoria tenuis*) to degrade the azo dyes by simple reduction. Living cells of *Chlorella vulgaris* remove 63–69 percent of the color from the mono-azo dye Tectilon Yellow 2G by transforming it to aniline (Aravindhan et al. 2007). Thus algae are frequently used for bioremediation of dye solution and textile wastewater. A number of studies on dye removal by algae are primarily based on using dead biomass which actually depends on biosorption rather than the degradation process.

2.2.3 Decolorization and Degradation of Dyes in Textile Effluent by Plants

Phytoremediation is the use of green plants or parts of them to detoxify toxic environmental pollutants. The use of a plant based detoxification system has several advantages such as it can be used *in situ* at the contaminated site, it is eco-friendly and low cost. Because of these advantages phytoremediation researchers are inclined to utilize them for the decolorization and degradation of dye and textile effluent. Sharma et al. (2005) used narrow leaved cattails *Typha angustifolia linn* for decolorization of synthetic reactive dye wastewater and achieved maximum color removal at day 14. Carias et al. (2008) worked on *Phragmites australis* which is a perennial wetland grass that effectively utilizes the ascorbate glutathione pathway for the detoxification of Acid Orange 7. Ong et al. (2009) constructed an up flow wetland reactor with an aeration system for the decolorization, degradation and mineralization of aromatic amines produced from the degradation of azo dye Acid Orange 7 containing wastewater. The wild plant *Blumea malcolmii* has been successfully used to decolorize Malachite Green (96.61 percent), Methyl Orange (87.86 percent), Red HE4B (76.50 percent), Reactive Red 2 (80.20 percent) and Direct Red 5B (42.07 percent) within three days. Among the above mentioned five different dyes, decolorization of Direct Red 5B was attributed to enzyme azoreductase, LiP and riboflavin reductase synthesized in plant roots (Kagalkar et al. 2009). In addition, Mbuligwe (2005) achieved 72–77 percent color removal in wetlands vegetated with Coco Yam plants of dye rich effluent. Furthermore, three plant species (*Brassica juncea*, *Sorghum vulgare* and *Phaseolus mungo*) have been studied for the decolorization of azo dyes and textile wastewater. These plants decolorized the textile wastewater up to the tune of 79 percent, 57 percent and 53 percent, respectively

(Ghodake et al. 2009b). Hairy root cultures of *Tagetes patula* L. (Marigold) have the potential to decolorize Reactive Red 198 and the polyphenol oxidase enzyme plays a crucial role in the oxidative degradation of dye (Patil et al. 2009). Shinde et al. (2012) demonstrated the potential of three fast growing plants *Parthenium hysterophorus*, *Alternanthera sessilis* and *Jatropha curcas* for decolorization of textile dyes Yellow 5G and Brown R mixture. The enzyme polyphenol oxidase extracted from these plants decolorizes dye formulation within six hours. Immobilized bitter gourd (*Momordica charantia*) peroxidase enzyme was used for the decolorization of textile effluent. The effective pH range 3.0–4.0 and temperature 40°C was found to be the optimum for enzyme mediated decolorization (Akhtar et al. 2005).

Phytoremediation blended with microorganisms has recently established its place among effective cleanup technology useful for the detoxification of xenobiotic dyes and converting them into nontoxic and less harmful products by synergy between plants and microbes (Tahir et al. 2017).

CHAPTER 3

Advanced Methods of Degradation of Dyes in Textile Effluent

T HE TEXTILE INDUSTRY IS an interdisciplinary area between natural and engineering sciences. It is continuously growing and coming up with new technologies to give us better quality and a range of products. Although the use of biological methods to treat the textile effluent has been a great success researchers are continuously trying to find new advanced methods for dye degradation which have additional benefits.

3.1 GENETICALLY MODIFIED MICROORGANISM AND THEIR ENZYMES

Biotechnology plays an important role in the development of more efficient and eco-friendly textile effluent treatment processes. It includes cloning, heterologous expression, gene recombination techniques, random mutagenesis, site directed mutagenesis, directed evolution, rational design and metagenomics to achieve an enhanced enhanced bioremediation process.

Furthermore, recent advances in molecular genetics and genetic engineering have led to cloning and expression of virtually any gene in a suitable microbial host (Figure 3.1). These processes can be useful for the absolute biodegradation of pollutants. The enzymes produced by cloning can be subjected to further rounds of random mutagenesis or gene recombination for different characteristics or alternative biochemical pathways in a microorganism. Although a single bacterium possesses superior decolorization capacity, its application in treating real wastewater is inadequate.

The knowledge of genetic engineering opens the door to creating genetically modified microorganisms (GMMs) with the required new metabolic pathways. GMMs are the alternative solution to wild strains which degrade dyes slowly if at all. The first GMM to be tested for degradation of naphthalene was *Pseudomonas fluorescence* HK44 in the USA for bioremediation

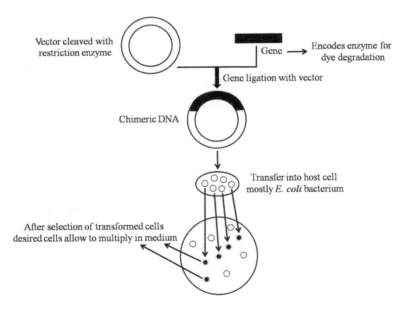

FIGURE 3.1 Basic steps in the production of genetically modified microorganisms for effective bio-removal of dyes

derivative) and develop into three dimensional tough assemblies which secrete a glue like substance that helps them to attach to any surface. The biofilm system is a promising technology that is useful in various contaminant removal processes such as biodegradation, biosorption, bioaccumulation and bio-mineralization (Pal et al. 2010). Qiu et al. (2013) designed an anaerobic upflow blanket bio-filter for the treatment of cationic red X-GRL dye containing wastewater. The filter contains proteobacteria including β-, γ-, δ- and ε-proteobacteria as the dominant microbes which primarily mediate decolorization and degradation of X-GRL. Anjaneya et al. (2013) demonstrated the potential of bacterial biofilms in the removal of dyes from industrial wastewater. These biofilms consist of strains AK1, AK2, VKY1 and a consortium that were used in batch, repeated batch and continuous packed bed bioreactor for the efficient decolorization of dye Amaranth.

Electrochemical active biofilms (EABs) have considerable attractions in the bio-electrochemical systems (BESs) field for pollutant degradation from wastewater. Similarly azo dye Amido Black 10B successfully decolorized in BES biofilms (Wang et al. 2016). Gong (2016) used four stage laboratory treatment systems, i.e. a moving bed biofilms reactor (MBBR) for the treatment of textile dyeing wastewater. The MBBR process improved the biodegradability of the raw wastewater and is also efficient in removing COD from dyeing wastewater. Harmer and Bishop (1992) investigated the removal of azo dye Acid Orange 7 from municipal wastewater by biofilms prepared from a variety of microorganisms found in activated sludge. Barragan et al. (2007) studied the effect of solid media such as bentonite and kaolin activated carbon on azo dye Acid Orange 7 biodegradation by three different bacteria (*Pseudomonas*, *Enterobacter* and *Morganella* sp.) under aerobic conditions. Sun et al. (2013) explored the degradation mechanism of Congo Red and bacterial diversity in a MFC (single chambered) that had a microfiltration membrane and air cathode.

Fungi based biofilms are also successfully used to degrade different synthetic azo dyes (including Reactive Red M-3BE,

Acid Red 249 and Reactive Black 5) and textile effluent with color removal efficiencies of 70–80 percent for 100 mg/L of dye solutions and 79–89 percent for textile wastewaters (Yang et al. 2004). Cerron et al. (2015) studied the decolorization potential of biofilms derived from *Trametes polyzona* LMB-TM5 and *Ceriporia* sp. LMB-TM1 isolated from the Peruvian Rainforest for six textile reactive dyes and effluents. Azo dyes were moderately decolorized by both strains whereas Synozol Turquoise Blue HF-G (phthalo-cyanine) and Remazol Brilliant Blue R (anthraquinone) were highly decolorized by *T. polyzona* LMB-TM5. Shabbir et al. (2017) utilized eco-friendly, green material, i.e. periphyton biofilm, for the bio-removal of an azo dye Methyl Orange and concluded that removal of dye attributed to a synergistic two stage mechanism involving bio-adsorption followed by biodegradation.

Biofilm mediated effluent treatment has several advantages such as operational flexibility, reduced hydraulic retention time, low space requirements, increased biomass residence time, enhanced ability to degrade recalcitrant compounds, high active biomass concentration and less production of sludge.

3.3 NANOPARTICLES

Recently the use of nanoparticles in the decolorization and degradation of textile dye has increased as it offers some advantages as it is eco-friendly in nature and more efficient than the other treatment methods available. Nanoparticles are the tiny structural arrangement with accuracy, intelligibility and less expensive. Various types of nanomaterials (Table 3.2) such as TiO_2, Zno and metallic nanoparticles (Au, Ag and Fe) have been tested for the catalytic degradation of textile dyes. The toxic organic dyes (Acid Orange, Malachite Green and Crystal Violet) were studied as models of textile effluent pollutant with the successful removal of dyes and other hazardous organic compounds showing the remediation ability of nanoparticles.

Over the past few years, numerous pathways have been discovered for the synthesis of transition metal oxide nanomaterials. The

TABLE 3.2 Nanoparticles used in the removal of dyes

Nanoparticles (NPs)	Dye(s)	References
Ag NPs	Acid Blue 113, Acid Black 24, Mordant Black 17	Sunkar and Vallinachiyar (2013)
Calcium Zincate NPs	Coralene Dark Red 2B	Yogendra et al. (2011)
CaMgO2 NPs	Brilliant Red	Gopalappa et al. (2014)
CaO and Tio2 NPs	Violet GL2B	Madhusudhana et al. (2012)
CdO Rhombus-like Nano Structure NPs	Crystal Violet, Malachite Green, Congo Red	Tadjarodi et al. (2014)
CdS and Ag/CdS NPs	DR-264	Fard et al. (2016)
Ferric Oxide NPs	Congo Red	Sultana et al. (2015)
Gold NPs	Methylene Blue	Narayanan and Park (2015)
Iron-Nickel/Metallic NPs	Orange G	Bokare et al. (2008)
Magnetic-Fe3O4 NPs	Reactive Blue	Darwesh et al. (2015)
NiFe2O4 NPs	Reactive Orange	Pak and Ardakani (2016)
Palladium/ Hydroxyapatite/ Fe3O4 NPs	Methyl Red, Methyl Orange, Methyl Yellow	Safavi and Momeni (2012)
Silver Nanoparticle NPs	Disperse Orange 3	Anuradha et al. (2015)
Template Cross linked-Chitosan NPs	Remazole Brilliant Orange, Remazol Black-5(RB-5)	Chen et al. (2011)
Tio2 NPs	CI Basic Blue 41	Giwa et al. (2012)
Triturated Copper NPs	Methyl Orange	Chakraborty et al. (2015)
Wo3 NPs	Congo Red, Rhodamine B	Mahshad et al. (2012)
Zn1-x Nixs and Zn1-x CuxS NPs	Congo Red	Pouretedal and Keshavarz (2011)
ZnO NPs	Brilliant Red	Jayamadhava et al. (2014)

Pseudomonas stutzeri AG259 strain, isolated from a silver mine was the first established proof that bacteria are involved in synthesizing silver nanoparticles. These Ag nanoparticles efficiently catalyze the degradation of Disperse Orange 3 and Methyl Orange in the presence of $NaBH_4$. Melanoidin decolorization using bacterial extract in an immobilized condition was also successfully achieved by using silver nanoparticles.

Iron nanoparticles are successfully synthesized by using green tea leaf and sorghum bran extracts (Nadagouda et al. 2010; Njagi et al. 2011). The synthesized iron nanoparticles encapsulated with alginate beads are utilized for removal of several toxic dyes, for example encapsulated iron nanoparticles using *Camellia sinensis* leaf extract. This is useful for degradation of Methyl Red and Methyl Orange dye. Iron oxides and zero valent iron (ZVI) combined with iron nanoparticles can be used as a Fenton-like catalyst for the removal of aqueous organic solutes (Xu et al. 2011). Similarly, Xu et al. (2008) employed Pd nanoparticles which had good catalytic properties for the removal of azo dyes. Cadmium oxide (CdO) is also used in the photodegradation method. CdO is an n-type semiconductor having direct and indirect band gap energy of 2.2–2.5 eV and 1.36–1.98 eV, respectively (Grado-Caffaro and Grado-Caffaro 2008).

Nano-bioremediation is a recently introduced technology to augment the process of bioremediation by using nanoparticles or nanotechnology. Chitosan, a natural aminopolysaccharide derivative of chitin is the most accepted and used adsorbent that removes dyes, metal ions and proteins from aqueous solutions. Acid Green 27 and Ecosin Y dyes are successfully bioadsorbed on chitosan based nanoparticles primed by ionic gelation with sodium tripolyphosphate (TPP) (Hu et al. 2006; Du et al. 2008).

Heterogeneous molecules are also exploited in the treatment of dye containing wastewater, for example the heterogeneous photocatalytic treatment is more efficient for the removal of organic dyes than the physical process as it involves the complete mineralization of organic compounds to CO_2, H_2O and mineral acids (Patrick and Dietmar 2007).

3.4 IMMOBILZATION

Immobilization is the process of physical confinement of intact cells/enzymes to a solid support or matrix with the maintenance of catalytic activity. Physical confinement is of two types, i.e. entrapment and attachment. Entrapment requires natural or synthetic polymers for immobilization. Natural polymers such as agar, chitosan, alginate, cellulose, collagen, gelatin, polyvinyl alcohol and synthetic polymers, such as a nylon or stainless steel sponge, polyurethane foam (PuF), have been used for the attachment procedure. The microbial cells as well as their enzymes have been successfully used in the decolorization and degradation of textile wastewater.

Paenibacillus alvei MTCC 10625 isolated from a textile wastewater contaminated site and immobilized over polyurethane foam (PUF) and nylon mesh (NM) showed the ability to decolorize synthetic dye effluent containing Black WNN dye up to 96.4 percent within 48 hours (Pokharia and Ahluwalia 2016b). Similarly another bacterial strain, *Stentrophomonas maltophilia,* was isolated from a polluted soil and immobilized on calcium alginate beads and showed a decolorization ability for the recalcitrant dye SITEX Black present in industrial effluent of up to 82 percent within 24 hours in batch conditions (Galai et al. 2010). Furthermore, a bacterial strain *Pseudomonas aeruginosa* immobilized on chitosan-magnetite nanoparticles demonstrated a potential for the quick decolorization of the azo dye Cibacron Red. The immobilized strain showed complete decolorization of both Cibacron Red dye and textile industrial wastewater after 2 hours and 12 hours for each treatment, respectively (Fetyan et al. 2017). Suganya and Revathi (2016) studied the competence of two sodium alginate and polyacrylamide gel bead immobilized strains of bacteria (*P. putida* and *B. licheniformis*) for the removal of reactive dyes (Reactive Red 195, Reactive Orange 72, Reactive Yellow 17, Reactive Blue 36) and found that sodium alginate immobilized bacteria achieved maximum decolorization of dyes.

Several immobilized strains of fungi also exhibited the potential to decolorize and degrade both dye and textile wastewater. Przystas et al. (2018) determined the capability of decolorization of two dyes of different classes, i.e. azo (Evans Blue) and triphenylmethane dyes (Brilliant Green) by immobilized fungal strains *Pleurotus ostreatus* BWPH, *Gleophyllum odoratum* DCa and *Polyporus picipes* RWP17. Although different solid supports are used for the immobilization of fungal strains, the most suit-able ones are brush, grid, washer and sawdust supports.

Vijayakumar and Manoharan (2012) investigated the efficiency of two Cyanobacteria, *Oscillatoria brevis* and *Westiellopsis prolific*, for industrial color wastewater treatment. The culture was studied both in free and immobilized conditions. These organisms not only removed the organic and inorganic content but also decreased the color intensity of wastewater by up to 75 percent within 30 days.

Al-Fawwaz and Abdullah (2016) studied the effects on decol-orization of Malachite Green and Methylene Blue by free and immobilized *Desmodesmus* sp. and achieved 98.6 percent decol-orization of both dyes with immobilized algae after six days. Similarly Kumar et al. (2014) revealed the ability of immobilized marine microalgae (*Chlorella marina, Tetraselmis* sp. *Dunaliella salina, Isochrysis galbana* and *Nannochloropsis* sp.) and freshwater microalga (*Chlorella* sp.) on 2 percent sodium alginate matrixes for the removal of dyes present in textile wastewater. Among the algal species studied the highest decolorization extent was shown by *Isochrysis galbana* (55 percent) followed by freshwater *Chlorella* sp. (43 percent).

Not only intact cells but also enzymes have been successfully immobilized and achieved the decolorization and degradation of textile effluent. Bilal and Asgher (2015) isolated MnP from a solid-state culture of fungi *G. lucidum* IBL-05 and immobilized it by using alginate beads. After the immobilization the catalytic potential of MnP was enhanced and it achieved 82.1 percent, 87.5 percent, 89.4 percent, 83 percent and 95.7 percent decolorization of Sandal-fix Turq Blue GWF, Sandal-fix Red C4BLN, Sandal-fix Foron Blue E2BLN, Sandal-fix Golden Yellow CRL and Sandal-fix Black CKF dyes, respectively. Furthermore, immobilized MnP also led to a significant reduction in biochemical oxygen demand (94.61–95.47 percent), chemical oxygen demand (91.18–94.85

percent) and total organic carbon (89.58–95 percent) of aqueous dye solutions. Plant derived polyphenol oxidase (PPO) enzyme extracted from *Cydonia oblonga* plant leaves was immobilized on calcium alginate beads and used for the effective decolorization of textile industrial wastewater (Arabaci and Usluoglu 2014).

3.5 MICROBIAL FUEL CELLS

In recent years, microbial fuel cells (MFCs) have emerged as a promising technology for the treatment of textile effluent. MFCs are considered to be a clean energy production tool as they can produce electricity. In MFCs, microorganisms work together with electrodes using electrons, which are either removed or supplied through an electrical circuit (Rabaey et al. 2007). The advantage associated with using MFCs is not only to fulfill increasing energy demand by using effluent as substrates to generate electricity, but also to accomplish effluent treatment simultaneously. Thus, it may reduce the costs of wastewater treatment. Besides producing electricity, an objective of this technology is also to treat pollutants such as nitrates, sulphide and sulphates present in wastewater.

It is a unique system which utilizes the chemical energy of biodegradable organic contaminant in effluent and converts it into biogenic electricity (MFCs) or into hydrogen/useful chemical products in microbial electrolysis cells (MECs) (Pant et al. 2012). In MFCs, the microbial population oxidizes organic matter (dye) in the anode chamber (anaerobic conditions) and generates electrons and protons. Electrons move to the cathode chamber via the external circuit where electrons, protons and the electron acceptor (mainly oxygen) combine to produce water (Li et al. 2009). The key components of MFCs are electrodes, substrate and membranes. Generally, the electrode materials possess some characteristics such as good chemical stability, good biocompatibility and high strength and efficiently transfer electrons between the bacterial and electrode surface. The MFC set up which is commonly used has either a single chamber or two chambers. In a two chamber set up, an ion-selective membrane separates the anode and cathode compartments which allow proton transfer from anode to cathode by preventing oxygen diffusion to the anode chamber. In the single chamber set up, the cathode is directly exposed to the air (oxygen). Besides these two common designs, several adaptations have been made in MFC design and structure to increase the efficiency of the system in textile wastewater treatment.

In the past few years, MFCs have been extensively studied as a new source of bio-energy and a tool for bioremediation. Debabov (2008) described a detailed mechanism of external electron transfer from two main bacteria (*Geobacter sulfurreducens* and *Shewanella oneidensis*) in a BES.

Several organisms that are used in MFCs have dye removal ability (Table 3.3). The bioelectricity generation through MFCs involves at least three mechanisms: (1) electron shuttle via cell secreting mediators (e.g. quinines, phenazine); (2) membrane associated redox proteins (e.g. cytochromes, a mobile electron carriers); and (3) conductive pili (or nanowires) (Chen et al. 2010b; Chen et al. 2012). The valuable aspect of MFC use is that the electrons generated in an MFC are utilized *in situ* for the removal of textile dyes and the electricity produced can be harvested from the system without an extra electricity supply. Electron transfer via cell secreting mediators plays a catalytic role in decreasing the activation energy and accelerates the oxidation rate of organic pollutants. Recent studies have revealed that some bacteria such as *Geobacter metallireducens*, *Shewanella putrefaciens*, *Geobacter sulfurreducens*, *Clostridium butyricum*, *Rhodoferax ferrireducens* and *Aeromonas hydrophila* transfer electrons directly to the electrodes without using mediators.

Electricity generation and the decolorization of azo dye using an MFC is largely influenced by many factors such as the nature of substrate, temperature and presence or absence of oxygen, etc. A complex substrate facilitates the establishment of a diverse and electrochemically active microbial community in the system while a simple substrate offers easier degradation and improves the electric and hydrogen output of the system.

TABLE 3.3   Dye degrading organisms used in the microbial fuel cell (MFC)

Dye/ Effluent	Microorganism	Decolorization (%)	Electricity generated	References
Active Brilliant Red X-3B	—	>100	234mW/m^2	Sun et al. (2009)
Amaranth	—	82.59	28.3W/m^2	Fu et al. (2010)
CI Acid Orange 7	—	97	5.0W/m^3	Zhang and Zhu (2011)
CI Reactive Blue 160	Acinetobacter johnsonii NIUx72, Enterobacter cancerogenus BYm30, Klebsiella pneumoniae ZMd31	—	200mV	Chen et al. (2010b)
Congo Red	—	98	103 mW/m^2	Cao et al. (2010)
Congo Red	—	69.3–92.7	387 mW/m^2	Li et al. 2008
Congo Red	—	—	324 mW/m2	Hou et al. (2011)
Dye wastewater	Proteus hauseri ZMd44	98	—	Chen et al. (2010c)
Methyl Orange		73.4	—	Ding et al. (2010)
Reactive Blue 221	—	83	305 Mv	Bakhshian et al. (2011)

MFCs at room or higher temperatures (20–35^{0}C) work relatively efficiently as compared to those at lower temperatures. MFCs decolorize azo dyes more efficiently under anaerobic (methanogenic) or anoxic conditions. Thus, these conditions are an operational strategy which offers cost effective bioremediation of pollutant such as dyes. During anaerobic decolorization, the accumulation of decolorized metabolites could significantly stimulate electron-transfer ability for dye degradation. However, model intermediate(s) with auxochromes (e.g. amino and hydroxyl substituted chemicals) could act as electron shuttles to feedback stimulation of MFC assisted degradation.

Effluent released from dye manufacturing and the textile industry contains a huge amount of synthetic dyes which are toxic in nature. Many research groups have successfully attempted to treat such dye wastewater by using MFCs. Sun et al. (2009) investigated the decolorization of an azo dye Active Brilliant Red X-3B (ABRX3) supplemented with readily biodegradable organic substrates such as glucose, sucrose, acetate and confectionery effluent by using a microfiltration membrane air cathode single chamber MFC. Biodegradation was found to be the primary dominant mechanism of the azo dye removal and glucose was used as the optimal co-substrate for complete removal of dye ABRX3 with maximum $234mW/m^2$ power density electricity generated. Cao et al. (2010) found up to 98 percent decolorization of azo dye Congo Red (300mg/L) in the anodic chamber of the MFC by using readily biodegradable co-substrates such as glucose, acetate and ethanol. Fu et al. (2010) demonstrated the degradation of azo dyes by the MFC-Fenton system which results in *in situ* production of H_2O_2 via two electron reduction of oxygen in a neutral catholyte. The MFC-Fenton system decolorized Amaranth (75mg/L) by 82.59 percent within one hour with maximum power density (PD) of $28.3W/m^2$. Chen et al. (2010c) studied the potential of indigenous *Proteus hauseri* ZMd44 for decolorization of dye amended wastewater in a single chamber MFC and achieved up to 98 percent removal of dye in 20 days. Further, Chen et al. (2010b) tested the ability of three indigenous dye decolorizers (e.g. *Acinetobacter johnsonii* NIUx72, *Klebsiella pneumoniae* ZMd31 and *Enterobacter cancerogenus* BYm30) against C.I. Reactive Blue 160 in air-cathode single chamber MFC and found that ZMd31 was involved in simultaneous dye decolorization and bioelectricity generation. Similarly, Zhang and Zhu (2011) tested the decolorization and degradation of azo dye Acid Orange 7 (AO7) containing wastewater in a single chamber MFC with simultaneous electricity generation. Furthermore, when glucose was used as co-substrate, nearly complete decolorization and degradation of AO7 was achieved after 168 hours of treatment with the maximum PD of $5.0W/m^3$. However, recently an aerobic biocathode MFC was used to treat the liquid having Active Brilliant Red X-3B (ABRX3) and 24.8 percent of COD removal was obtained from the decolorization liquid of ABRX3 without using glucose as a co-substrate by the biocathode within 12 hours (Sun et al. 2011). Bakhshian et al. (2011) investigated the role of the enzyme in decolorization of Reactive Blue 221 (RB221) by using laccase in the cathode chamber and molasses in the anode chamber of a dual chamber MFC, which resulted in 83 percent dye decolorization and 84 percent COD removal from molasses in the cathode and anode chamber, respectively.

3.6 CONCLUSION AND FUTURE PROSPECTIVE

Effluent released by the textile industries is a complex mixture of various pollutants including a variety of textile dyes. The discharge of dye containing textile effluent into the environment has caused serious concern because of its aesthetic value, toxicity, mutagenicity and carcinogenicity. Various physico-chemical methods have been used for the treatment of textile effluent but these methods have several limitations. Biological treatment methods based on microorganisms (bacteria, fungi, algae) and plants are an effective and environmentally friendly alternative for the removal of textile dyes and other pollutants present in textile effluent. The microbial decolorization of textile dyes can occur by either biosorption or enzymatic degradation or by a combination of both. The microorganisms are capable of producing enzymes (oxidative and reductive) which catalyze the breakdown of textile dyes. Treatment systems based on consortia are more effective as compared to a single pure microorganism due to the synergistic metabolic activities of the microbial community. Several researchers are also focusing on the isolation of mixed microbial communities or consortia that have the potential to remove the textile dyes and tolerate the extreme conditions of textile effluent. Various advanced methods such as GMMs and their enzymes, immobilized cells or enzymes, biofilms and MFCs have great potential for the treatment of textile effluent. However, the potential of microbes and other advanced methods have been tested using dye solution under lab conditions only in most cases. The ability of these methods to treat real textile effluent has not been fully elucidated. Therefore, it will be necessary to evaluate the potential of capable microbes and other advanced methods to treat the real textile effluent for the development of an effective treatment technology. The isolation of new microbial strains or the acclimatization of existing ones for the removal of textile dyes will probably increase the efficacy of the biological treatment system in the near future. Moreover, the metabolic pathways involved in microbial degradation of textile effluent have not yet been fully explained. Hence, future work should focus on underlying processes and metabolic pathways by identifying the genes and metabolites in the microbial degradation processes. Textile industries consume large amounts of potable water and, therefore, recycling and reuse of textile effluent is also necessary to limit the amount of wastewater and protect the environment. Treatment systems having the capability to complete the mineralization of textile dyes and other pollutants present in textile effluent as well as to recycle and reuse the treated water are going to be necessary in the future.

References

Abraham, K. J., and G. H. John. 2007. Development of a classification scheme using a secondary and tertiary amino acid analysis of azoreductase gene. *Journal of Medical and Biological Sciences* 1 (1): 1–5.

Acuner, E., and F. B. Dilek. 2004. Treatment of Tectilon Yellow 2G by *Chlorella vulgaris*. *Process Biochemistry* 39: 623–631.

Akhtar, S., A. A. Khan, and Q. Husain. 2005. Potential of immobilized bitter gourd (*Momordica charantia*) peroxidases in the decolorization and removal of textile dyes from polluted wastewater and dyeing effluent. *Chemosphere* 60: 291–301.

Al-Fawwaz, A. T., and M. Abdullah. 2016. Decolorization of Methylene Blue and Malachite Green by immobilized *Desmodesmus* sp. isolated from North Jordan. International Journal of Environmental Science and Development 7 (2): 95–99.

Andleeb, S. N., M. I. Atiq, R. Ali, M. Razi-Ul-Hussnain, B. Shafique, P. B. Ahmad, M. Ghumro et al. 2010. Bological treatment of textile effluent in stirred tank bioreactor. *International Journal of Agriculture and Biology* 12: 256–260.

Anjaneya, O., S. S. Shrishailnath, G. K. Anand, S. Nayak, S. B. Mashetty, and T. B. Karegoudar. 2013. Decolorization of Amaranth dye by bacterial biofilm in batch and continuous packed bed bioreactor. *International Biodeterioration & Biodegradation* 79: 64–72.

Anuradha, G. M., K. Yadav, and N. Jaggi. 2015. Degradation of Disperse Orange 3 an azo dye by silver NPs. *Indian Journal of Chemical Technology* 22: 167–170.

Apostol, L. C., L. Pereira, R. Pereira, M. Gavrilescu, and M. M. Alves. 2012. Biological decolorizartion of Xanthene dyes by anaerobic granular biomass. *Biodegradation* 23 (5): 725–737. doi: 10.1007/s10532-012-9548-7.

Arabaci, G., and A. Usluoglu. 2014. The enzymatic decolorization of textile dyes by the immobilized polyphenol oxidase from quince leaves. *Scientific World Journal* doi.org/10.1155/2014/685975.

Aravindhan, R., J. R. Rao, and B. U. Nair. 2007. Removal of basic yellow dye from aqueous solution by sorption on green alga *C. pelliformis*. *Journal of Hazardous Materials* 142: 68–76.

Arora, D. S., and R. K. Sharma. 2010. Ligninolytic fungal laccases and their biotechnological applications. *Applied Biochemistry and Biotechnology* 160: 1760–1788.

Ayca, J., U. Nilsson, E. Terrazas, T. A. Aliaga, and U. Welander. 2006. Decolorization of the textile dyes using cyanobacterium in a continuous rotating biological contactor reactor. *Enzyme and Microbial Technology* 39: 32–37.

Ayed, L., A. Mahdhi, A. Cheref, and A. Bakhrouf. 2011. Decolorization and degradation of azo dye Methyl Red by an isolated *Sphingomonas paucimobilis*: biotoxicity and metabolites characterization. *Desalination* 274: 272–277.

Azza, M. A., S. A. Nabila, A. G. Hany, and K. A. Rizka. 2013. Biosorption of cadmium and lead from aqueous solution by fresh water alga *Anabaena sphaerica* biomass. *Journal of Advanced Research* 4: 367–374.

Babu, B. R., A. K. Parande, S. Raghu, and T. P. Kumar. 2007. Cotton textile processing: Waste generation and effluent treatment. *The Journal of Cotton Science* 11: 141–153.

Bakhshian, S., H. R. Kariminia, and R. Roshandel. 2011. Bioelectricity generation in a dual microbial fuel cell under cathodic enzyme catalyzed dye decolorization. *Bioresource Technology* 102 (12): 6761–6765.

Barragan, B. E., C. Costa, and M. C. Marquez. 2007. Biodegradation of azo dyes by bacteria inoculated on solid media. *Dyes and Pigments* 75: 73–81.

Bergsten-Torralba, L. R., M. M. Nishikawa, D. F. Baptista, D. P. Magalhaes, and M. da Silva. 2009. Decolorization of different textile dyes by *Penicillium simplicissimum* and toxicity evaluation after fungal treatment. *Brazilian Journal of Microbiology* 40: 808–817.

Bilal, M., and M. Asgher. 2015. Dye decolorization and detoxification potential of Ca-alginate beads immobilized manganese peroxidase. *BMC Biotechnology* 15: 111. doi.org/10.1186/s12896-015-0227-8.

Bin, Y., Z. Jiti, W. Jing, D. Cuihong, H. Hongman, S. Zhiyong, and B. Yongming. 2004. Expression and characteristics of the gene encoding azoreductase from *Rhodobacter sphaeroides* AS1.1737. *FEMS Microbiology Letters* 236: 129–136.

Blumel, S., and Stolz, A. 2003. Cloning and characterization of the gene coding for the aerobic azoreductase from *Pigmentiphaga kullae* K24. *Applied Microbiology and Biotechnology* 62: 186–190.

Blumel, S., H. J. Knackmuss, and A. Stolz. 2002. Molecular cloning and characterization of the gene coding for the aerobic azoreductase from *Xenophilus azovorans* KF46F. *Applied and Environmental Microbiology* 68 (8): 3948–3955.

Bokare, A. D., R. C. Chikate, V. C. Rode, and K. M. Paknikar. 2008. Iron nickel bimetallic nanoparticles for reductive degradation of azo dye Orange G in aqueous solution. *Applied Catalysis B: Environmental* 79: 270–278.

Bragger, J. L., A. W. Lloyd, S. H. Soozandehfar, S. F. Bloomfield, C. Marriott, and G. P. Martin. 1997. Investigations into the azo reducing activity of a common colonic microorganism. *International Journal of Pharmaceutics* 157: 61–71.

Cao, Y., Y. Hu, J. Sun, and B. Hou. 2010. Explore various co-substrates for simultaneous electricity generation and Congo Red degradation in air cathode single chamber microbial fuel cell. *Bioelectrochemistry* 79: 71–76.

Carias, C. C., J. M. Novais, and S. M. Dias. 2008. Are *Phragmites australis* enzymes involved in the degradation of the textile azo dye Acid Orange 7? *Bioresource Technology* 99 (2): 243–251.

Carneiro, P. A., G. A. Umbuzeiro, D. P. Oliveira, and M. V. B. Zanoni. 2010. Assessment of water contamination caused by a mutagenic textile effluent/dyehouse effluent bearing Disperse dyes. *Journal of Hazardous Materials* 174: 694–699.

Cerron, L., D. Romero-Suarez, N. Vera, and Y. Ludena. 2015. Decolorization of textile reactive dyes and effluents by biofilms of *Trametes polyzona* LMB-TM5 and *Ceriporia* sp. LMB-TM1 isolated from the Peruvian rainforest. *Water Air and Soil Pollution* 226 (7): 235 (1–13); doi.org/10.1007/s11270-015-2505-4.

Chacko, J. T., and K. Subramaniam. 2011. Enzymatic degradation of azo dyes-a review. *International Journal of Environmental Sciences* 1 (6): 1250–1260.

Chakraborty, M., R. Basu, S. Das, and P. Nandy. 2015. Application of triturated copper nanoparticles as an agent for remediation of an azo dye Methyl Orange. *MOL2NET* 1: 1–7.

Chang, J., and C. Lin. 2001. Decolorization kinetics of a recombinant *Escherichia coli* strain harboring azo dye decolorizing determinants from *Rhodococcus* sp. *Biotechnology Letters* 23 (8): 631–636.

Charumathi, D., and N. Das. 2010. Bioaccumulation of synthetic dyes by *Candida tropicalis* growing in sugarcane bagasse extract medium. *Advances in Biological Research* 4 (4): 233–240.

Chen, B. Y., M. M. Zhang, Y. Ding, and C. T. Chang. 2010b. Feasibility study of simultaneous bioelectricity generation and dye decolorization using naturally occurring decolorizes. *Journal of Taiwan Institute of Chemical Engineers* 41: 682–688.

Chen, B. Y., M. M. Zhang, C. T. Chang, Y. Ding, K. L. Lin, C. S. Chiou, C. C. Hsueh, and H. Xu. 2010c. Assessment upon azo dye decolorization and bioelectricity generation by *Proteus hauseri*. *Bioresource Technology* 101: 4737–4741.

Chen, B. Y., Y. M. Wan, I. S. Ng, S. Q. Liu, and J. Y. Hung. 2012. Deciphering simultaneous bioelectricity generation and dye decolorization using *Proteus hauseri*. *Journal of Bioscience and Bioengineering* 113: 502–507.

Chen, C. Y., J. C. Chang, and A. H. Chen. 2011. Competitive biosorption of azo dyes from aqueous solution on the templated cross linked-chitosan nanoparticles. *Journal of Hazardous Materials* 185: 430–441.

Chen, H., J. Feng, O. Kweon, H. Xu, and C. E. Cerniglia. 2010a. Identification and molecular characterization of a novel flavin-free NADPH preferred azoreductase encoded by *azoB* in *Pigmentiphaga kullae* K24. *BMC Biochemistry* 11: 13. doi: 10.1186/1471-2091-11-13.

Chen, H., R. F. Wang, and C. E. Cerniglia, 2004. Molecular cloning, overexpression, purification, and characterization of an aerobic FMN-dependent azoreductase from *Enterococcus faecalis*. *Protein Expression and Purification* 34: 302–310.

Chen, H., S. L. Hopper, and C. E. Cerniglia. 2005. Biochemical and molecular characterization of an azoreductase from *Staphylococcus aereus*. A tetrameric NADPH—dependent flavoprotein. *Microbiology* 151: 1433–1441.

Chen, K. C., J. Y. Wua, D. J. Liou, and S. J. Hwang. 2003. Decolorization of the textile dyes by newly isolated bacterial strains. *Journal of Biotechnology* 101: 57–68.

Cheriaa, J., M. Khaireddine, M. Rouabhia, and A. Bakhrouf. 2012. Removal of triphenylmethane dyes by bacterial consortium. *Scientific World Journal*, Article ID 512454. doi: 10.1100/2012/512454.

Chivukula, M., J. T. Spadaro, V. Renganathan. 1995. Lignin peroxidase-catalyzed oxidation of sulfonated azo dyes generates novel sulfophenyl hydroperoxides. *Biochemistry* 34: 7765–7772.

Clark, M. 2011. *Handbook of Textile and Industrial Dyeing. Volume 1 Principles, Processes and Types of Dyes.* Woodhead Publishing Series in Textiles.

Colao, M. C., S. Lupino, A. M. Garzillo, V. Buonocore, and M. Ruzzi. 2006. Heterologous expression of lccl gene from *Trametes trogii* in *Pichia pastoris* and characterization of the recombinant enzyme. *Microbial Cell Factories* 5: 31–36.

Dale, J. W., and S. F. Park. 2007. *Molecular Genetics of Bacteria.* Chichester: John Wiley & Sons, UK, pp. 137–244.

Daneshvar, N., M. Ayazloo, A. R. Khataee, and M. Pourhassan. 2007. Biological decolorization of dye solution containing Malachite Green by microalgae *Cosmarium* sp. *Bioresource Technology* 98: 1176–1182.

Darwesh, O. M., H. Moawad, O. S. Barakat, and W. M. A. El-Rahim. 2015. Bioremediation of textile reactive blue azo dye residues using nanobiotechnology approaches. *Research Journal of Pharmaceutical, Biological and Chemical Sciences* 6 (1): 1202–1211.

Dawkar, V. V., U. U. Jadhav, G. S. Ghodake, and S. P. Govindwar. 2009. Effect of inducers on the decolorization and biodegradation of textile azo dye Navy Blue 2GL by *Bacillus* sp. VUS. *Biodegradation* 20 (6): 777–787. doi: 10.1007/s10532-009-9266-y.

De Roos, A. J., R. M. Ray, D. L. Gao, K. J. Wernli, E. D. Fitzgibbons, F. Ziding, G. Astrakianakis, D. B. Thoma, and H. Checkoway. 2005. Colorectal cancer incidence among female textile workers in Shanghai, China: A case-cohort analysis of occupational exposures. *Cancer Causes and Control* 16 (10): 1177–1188.

Debabov, V. G. 2008. Electricity from microorganisms. *Microbiologia* 77 (2): 149–157.

Deller, S., S. Sollner, R. Trenker-El-Toukhy, I. Jelesarov, G. M. Gubitz, and P. Macheroux. 2006. Characterization of a thermostable NADPH:FMN oxidoreductase from the mesophilic bacterium *Bacillus subtilis. Biochemistry* 45: 7083–7091.

Deng, D., J. Guo, G. Zeng, and G. Sun. 2008. Decolorization of anthraquinone, triphenylmethane and azo dyes by a new isolated *Bacillus cereus* strain DC11. *International Biodeterioration & Biodegradation* 62: 263–269.

Dhanve, R. S., U. U. Shedbalkar, and J. P. Jadhav. 2008. Biodegradation of diazo reactive dye Navy blue HE2R (reactive blue 172) by an isolated *Exiguobacterium* sp. RD3. *Biotechnology and Bioprocess Engineering* 13: 53–60.

Ding, H., Y. Li, A. Lu, S. Jin, C. Quan, C. Wang, X. Wang, C. Zeng, and Y. Yan. 2010. Photocatalytically improved azo dye reduction in a microbial fuel cell with rutile cathode. *Bioresource Technology* 101: 3500–3505.

Dixit, S., and S. Garg. 2018. Biodegradation of environmentally hazardous azo dyes and aromatic amines using *Klebsiella pneumonia*. *Journal of Environmental Engineering* 144 (6). doi: 10.1061/(ASCE)EE.1943-7870.0001353.

Dorthy, C. A. M., R. Sivarj, and R. Venckatesh. 2012. Decolorization efficiencies of dyes and effluent by free and immobilized fungal isolates. *International Journal of Environmental Science Research* 1 (4): 109–113.

Du, W. L., R. Xu, X. Y. Han, Y. L. Xu, and Z. G. Miao. 2008. Preparation characterization and adsorption properties of chitosan nanoparticles for Eosin Y as a model anionic dye. *Journal of Hazardous Material* 153: 152–156.

Duarte-Vazquez, M. A., J. R. Whitaker, A. Rojo-Dominguez, B. E. Garcia-Almendarez and, C. Regalado. 2003. Isolation and thermal characterization of an acidic iso-peroxidase from turnip roots. *Journal of Agriculture and Food Chemistry* 51: 5096–5102. doi:10.1021/jf026151y.

Dube, E., F. Shareck, Y. Hurtubise, C. Daneault, and M. Beauregard. 2008. Homologous cloning, expression, and characterization of a laccase from *Streptomyces coelicolor* and enzymatic decolorization of an Indigo dye. *Applied Microbiology and Biotechnology* 79 (4): 597–603.

Duran, N., M. A. Rosa, A. D'Annibale, L. Gianfreda. 2002. Applications of laccases and tyrosinases (phenoloxidases) immobilized on different supports: A review. *Enzyme and Microbial Technology* 31: 907–931.

Erickson, R. J., and J. M. McKim. 1990. A simple flow-limited model for exchange of organic chemicals at fish gills. *Environmental Toxicology and Chemistry* 9: 159–165.

Ertugrul, S., M. Bakira, and G. Donmez. 2008. Treatment of dye rich wastewater by an immobilized thermophilic cyanobacterial strain: *Phormidium* sp. *Ecological Engineering* 32: 244–248.

Esancy, J. F., H. S. Freeman, L. D. Claxton. 1990. The effect of alkoxy substituents on the mutagenicity of some aminoazobenzene dyes and their reductive-cleavage products. *Mutation Research* 238 (1): 1–22.

Fard, N. E., R. Fazaeli, and R. Ghiasi. 2016. Band gap energies and photocatalytic properties of Cds and Ag/Cds NPs for azo dye degradation. *Chemical Engineering & Technology* 39 (1): 149–157.

Fetyan, N. A. H., A. Z. A. Azeiz, I. M. Ismail, and T. M. Salem. 2017. Biodegradation of Cibacron Red azo dye and industrial textile effluent by *Pseudomonas aeruginosa* immobilized on chitosan-Fe_2O_3 composite. *Journal of Advances in Biology & Biotechnology* 12 (2): 1–15.

Flores, E. R., M. Luijten, and B. A. Donlon et al. 1997. Complete biodegradation of the azo dye azodisalicylate under anaerobic condition. *Environmental Science & Technology* 31: 2098–2103.

Fu, L., S. J. You, G. Q. Zhang, F. L. Yang, and X. H. Fang. 2010. Degradation of azo dyes using *in situ* Fenton reaction incorporated into H_2O_2 producing microbial fuel cell. *Chemical Engineering Journal* 160: 164–169.

Fu, Y., and T. Viraraghavan. 2001. Fungal decolorization of dye wastewaters: A review. *Bioresource Technology* 79: 251–262.

Gahr, F., F. Hermanutz, and W. Oppermann. 1994. Ozonation an important technique to comply with new German laws for textile wastewater treatment. *Water Science & Technology* 30: 255–263.

Galai, S., F. Limam, and M. N. Marzouki. 2010. Decolorization of an industrial effluent by free and immobilized cells of *Stenotrophomonas maltophilia* AAP56. Implementation of efficient down flow column reactor. *World Journal of Microbiology and Biotechnology* 26: 1341–1347.

Ghodake, G., S. Jadhav, V. Dawkar, and S. Govindwar. 2009a. Biodegradation of diazo dye Direct Brown MR by *Acinetobacter calcoaceticus* NCIM 2890. *International Biodeterioration & Biodegradation* 63: 433–439.

Ghodake, G. S., A. A. Telke, J. P. Jadhav, and S. P. Govindwar. 2009b. Potential of *Brassica juncea* in order to treat textile effluent contaminated sites. *International Journal of Phytoremediation* 11: 297–312.

Giardina, P., V. Faraco, C. Pezzella, A. Piscitelli, S. Vanhulle, and G. Sannia. 2010. Laccases: A never-ending story. *Cellular and Molecular Life Sciences* 67: 369–385.

Gingell, R., J. W. Bridges, and R. T. Williams. 1971. The role of the gut flora in the metabolism of *Prontosil* and *Neoprontosil* in the Rat. *Xenobiotica* 1 (2): 143–156.

Giwa, A., P. O. Nkeonye, B. A. Bello, and K. A. Kolawole. 2012. Photocatalytic decolorization and degradation of C.I. Basic Blue 41 using TiO2 nanoparticles. *Journal of Environmental Protection* 3 (9): 1063–1069.

Gong, J. B. 2016. Advanced treatment of textile dyeing wastewater through the combination of moving bed biofilm reactors and ozonation. *Separation Science and Technology* 51 (9): 1589–1597. doi: 10.1080/01496395.2016.1165703.

Gopalappa, H., Yogendra, K. M. Mahadevan, N. Madhusudhana. 2014. Solar photocatalytic degradation of azo dye Brilliant Red in aqueous medium by synthesized $CaMgO_2$ nanopartices as an alternative catalyst. *Chemical Science Transactions* 3 (1): 232–239.

Grado-Caffaro, M. A., and M. Grado-Caffaro. 2008. A quantitative discussion on band gap energy and carrier density of CdO in terms of temperature and oxygen partial pressure. *Physics Letter A* 372: 4858–4860.

Grassi, E., P. Scodeller, N. Filiel, R. Carballo, and L. Levin. 2011. Potential of *Trametes trogii* culture fluids and its purified laccase for the decolorization of different types of recalcitrant dyes without the addition of redox mediators. *International Biodeterioration & Biodegradation* 65: 635–643. doi: 10.1016/j.ibiod.2011.03.007.

Greene, J. C. and G. L. Baughman. 1996. Effects of 46 dyes on population growth of fresh water green alga *Selenastrum capricornutum*. *Textile Chemist and Colorist* 28: 23–30.

Guan, Z. B., S. Yan, C. M. Song, N. Zhang, Y. J. Cai, and X. R. Liao. 2015. Efficient secretary production of CotA-laccase and its application in the decolorization and detoxification of industrial textile wastewater. *Environment Science and Pollution Research* 22: 9515–9523.

Guolan, H., S. Hongwen, and L. L. Cong. 2000. Study on the physiology and degradation of dye with immobilized algae. *Artificial Cells Blood Substitutes and Biotechnology* 28: 347–363.

Gurav, A. A., J. S. Ghosh, and G. S. Kulkarni. 2011. Decolorization of anthroquinone based dye Vat Red 10 by *Pseudomonas desmolyticum* NCIM 2112 and *Galactomyces geotrichum* MTCC 1360. *International Journal for Biotechnology and Molecular Biology Research* 2 (6): 93–97.

Hao, O. J., K. Hyunook, and P. Chiang. 2000. Decolorization of wastewater. *Critical Reviews in Environmental Science and Technology* 30: 449–505.

Harmer, C., and P. Bishop. 1992. Transformation of azo dye AO 7 by wastewater biofilms. *Water Science and Technology* 26 (3–4): 627–636.

He, X. L., C. Song, Y. Y. Li, N. Wang, L. Xu, X. Han, and D. S. Wei. 2018. Efficient degradation of Azo dyes by a newly isolated fungus *Trichoderma tomentosum* under non-sterile conditions. *Ecotoxicology and Environmental Safety* 150: 232–239.

Hong, Y., J. Guo, and G. Sun. 2009. Energy generation coupled to the azoreduction by the membranous vesicles from *Shewanella decolorationis* S12. *Journal of Microbiology and Biotechnology* 19: 37–41.

Hou, B., J. Sun, and Y. Y. Hu. 2011. Simultaneous Congo Red decolorization and electricity generation in air cathode single chamber microbial fuel cell with different microfiltration, ultrafiltration and proton exchange membrane. *Bioresource Technology* 102 (6): 4433–4438.

Hu, Z. G., J. Zhang, W. L. Chan, Chan, Szeto, and Y. S. Szeto. 2006. The sorption of acid dye onto chitosan nanoparticles. *Polymer* 47: 5838–5842.

Huang, Y. H., C. L. Hsueh, C. P. Huang, L. C. Su, and C. Y. Chen. 2007. Adsorption thermodynamic and kinetic studies of Pb(II) removal from water onto a versatile Al_2O_3-supported iron oxide. *Separation Purification Technology* 55: 23–29.

Ince, N. H., and G. Tezcanli. 1999. Treatability of textile dye-bath effluents by advanced oxidation: Preparation for reuse. *Water Science & Technology* 40: 183–190.

Ito, T., Y. Shimada, and T. Suto. 2018. Potential use of bacteria collected from human hands for textile dye decolorization. *Water Resources and Industry* 20: 46–53.

Jadhav J. P., D. C. Kalyani, A. A. Telke, S. S. Phugare, and S. P. Govindwar. 2010. Evaluation of the efficacy of a bacterial consortium for the removal of color, reduction of heavy metals and toxicity from textile dye effluent. *Bioresource Technology* 101: 165–173.

Jadhav, S. B., S. S. Phugare, P. S. Patil, and J. P. Jadhav. 2011. Biochemical degradation pathway of textile dye Remazol Red and subsequent toxicological evaluation by cytotoxicity, genotoxicity and oxidative stress studies. *International Biodeterioration & Biodegradation* 65: 733–743.

Jadhav, S. U., M. U. Jadhav, A. N. Kagalkar, and S. P. Govindwar. 2008. Decolorization of Brilliant Blue G dye mediated by degradation of the microbial consortium of *Galactomyces geotrichum* and *Bacillus* sp. *Journal of the Chinese Institute of Chemical Engineers* 39: 563–570.

Jayamadhava, P., A. Sudhakara, S. Ramesha, and G. Nataraja. 2014. Synthesize of ZnO NPs as an alternative catalyst for photocatalytic degradation of Brilliant Red azo dye. *American Journal of Environmental Protection* 3 (6): 318–322.

Jin, R., H. Yang, A. Zhang, J. Wang, and G. Liu. 2009. Bioaugmentation on decolorization of C.I. Direct Blue 71 by using genetically engineered strain *Escherichia coli* JM109 (PGEX-AZR). *Journal of Hazardous Materials* 163: 1123–1128.

Kagalkar, A. N., U. B. Jagtap, J. P. Jadhav, V. Bapat, and S. P. Govindwar. 2009. Biotechnological strategies for phytoremediation of the sulfonated azo dye Direct Red 5B using *Blumea malcolmii* hook. *Bioresource Technology* 100: 4104–4110.

Khataee, A. R., G. Dehghan, A. Ebadi, M. Zarei, and M. Pourhassan. 2010. Biological treatment of a dye solution by Macroalgae *Chara* sp: Effect of operational parameters, intermediates identification and artificial neural network modelling. *Bioresource Technology* 101: 2252–2258.

Khehra, M. S., H. S. Saini, D. K. Sharma, B. S. Chadha, and S. S. Chimni. 2006. Biodegradation of azo dye C.I. Acid Red 88 by anoxic-aerobic sequential bioreactor. *Dyes and Pigments.* 70: 1–7.

Khehra, M. S., H. S. Sainia, D. K. Sharma, B. S. Chadha, and S. S. Chimni. 2005. Decolorization of various azo dyes by bacterial consortium. *Dyes and Pigments* 67: 55–61.

Khlifi, R., L. Belbahri, S. Woodward, M. Ellouz, A. Dhouib, S. Sayadi, T. Mechichi. 2010. Decolorization and detoxification of textile industry wastewater by the laccase-mediator system. *Journal of Hazardous Materials* 175: 802–808.

Kim, S. J., and P. Shoda. 1999. Purification and characterization of a novel peroxidase from *Geotrichum candidum* dec 1 involved in decolorization of dyes. *Applied and Environmental Microbiology* 65: 1029–1035.

Kirby, N. 1999. Bioremediation of textile industry wastewater by white rot fungi. DPhil Thesis, University of Ulster, Coleraine, UK.

Kirby, N., R. Marchan, and G. McMullan. 2000. Decolourization of synthetic textile dyes by *Phlebia tremellosa*. *FEMS Microbiology Letters* 188: 93–96.

Kodam, K. M., I. Soojhawon, P. D. Lokhande, and K. R. Gawai. 2005. Microbial decolorization of reactive azo dyes under aerobic conditions. *World Journal of Microbiology and Biotechnology* 21: 367–370.

Kuberan, T., J. Anburaj, C. Sundaravadivelan, and P. Kumar. 2011. Biodegradation of azo dye by *Listeria* sp. *International Journal of Environmental Sciences* 1 (7): 1760–1770.

Kumar, D. S., P. Santhanam, K. R. Nanda, S. Ananth, P. B. Balaji, S. A. Devi, S. Jeyanthi, and P. Ananthi. 2014. Preliminary study on the dye removal efficiency of immobilized marine and freshwater microalgal beads from textile wastewater. *African Journal of Biotechnology* 13 (22): 2288–2294.

Kumar, M. N. V. R., T. R. Sridhar, K. D. Bhavani, and P. K. Dutta. 1998. Trends in color removal from textile mill effluents. *Colorage* 40: 25–34.

Lee, Y. C., and S. P. Chang. 2011. The biosorption of heavy metals from aqueous solution by *Spirogyra* and *Cladophora* filamentous macroalgae. *Bioresource Technology* 102: 5297–5304.

Li, J. F., Y. Z. Hong, and Y. Z. Xiao. 2007. Cloning and heterologous expression of the gene of laccase C from *Trametes* sp. 420 and potential of recombinant laccase in dye decolorization. *Acta Microbiologica Sinica* 47: 54–58.

Li, Z., X. Zhang, J. Lin, S. Han, and L. Lei. 2008. Azo dye treatment with simultaneous electricity production in an anaerobic-aerobic sequential reactor and microbial fuel cell coupled system. *Bioresource Technology* 101 (12): 4440–4445.

Li, Z., X. Zhang, Y. Zeng, and L. Lei. 2009. Electricity production by an overflow type wetted microbial fuel cell. *Bioresource Technology* 100: 2551–2555.

Lim, S. L., W. L. Chu, and S. M. Phang. 2010. Use of *Chlorella vulgaris* for bioremediation of textile wastewater. *Bioresource Technology* 101: 7314–7322.

Lima, R. O. A., A. P. Bazo, D. M. Salvadori, C. M. Rech, D. P. Oliveira, and G. A. Umbuzeiro. 2007. Mutagenic and carcinogenic potential of a textile azo dye processing plant effluent that impacts a drinking water source. *Mutation Research* 626 (1–2): 53–60.

Liu, J. Q., and H. T. Liu. 1992. Degradation of azo dyes by algae. *Environmental Pollution* 75: 273–278.

Lu, I., M. Zhao, S. C. Liang, L. Y. Zhao, D. B. Li, and B. B. Zhang. 2009. Production and synthetic dyes decolorization capacity of a recombinant laccase from *Pichia pastoris*. *Journal of Applied Microbiology* 107: 1149–1156.

Macwana, S. R., S. Punj, J. Cooper, E. Schwenk, G. H. John. 2010. Identification and isolation of an azoreductase from *Enterococcus faecium*. *Current Issues in Molecular Biology* 12: 43–48.

Madhusudhana, N., K. Yogendra, and K. M. Mahadevan. 2012. A comparative study on photocatalytic degradation of Violet GLB2B azo dye using CaO and TiO2 NPs. *International Journal of Engineering Research and Application* 2 (5): 1300–1307.

Mahshad, A., M. A. Reza, and R. Alimorad. 2012. Effect of WO_3 nanoparticles on Congo Red and Rhodamine-B photodegradation. *Iranian Journal of Chemistry & Chemical Engineering* 31 (1): 23–29.

Manu, B., and S. Chaudhari. 2002. Anaerobic decolorization of simulated textile wastewater containing azo dyes. *Bioresource Technology* 82: 225–231.

Martins, L. O., C. M. Soares, M. M. Pereira et al. 2002. Molecular and biochemical characterization of a highly stable bacterial laccase that occurs as a structural component of the *Bacillus subtilis* endospore coat. *Journal of Biological Chemistry* 277: 18849–18859.

Masoud, K., D. Ehsan, D. Hakimeh, T. Delaram, and B. Amit. 2012. Box-Behnken design optimization of Acid Black 1 dye biosorption by different brown macroalgae. *Chemical Engineering Journal* 179: 158–168.

Matsumoto, K., Y. Mukai, D. Ogata, F. Shozui, J. M. Nduko, S. Taguchi, and T. Ooi 2010. Characterization of thermostable FMN-dependent NADH azoreductase from the moderate thermophile *Geobacillus stearothermophilus*. *Applied Microbiology Biotechnology* 86: 1431–1438.

Mbuligwe, S. E., 2005. Comparative treatment of dye-rich wastewater in engineered wetland systems (EWSs) vegetated with different plants. *Water Research* 39: 271–280.

Mendes, S., L. Pereira, C. Batista, and L. O. Martins. 2011. Molecular determinants of azo reduction activity in the strain *Pseudomonas putida* MET94. *Applied Microbiology & Biotechnology* 92: 393–405.

Modi, H. A., G. Rajput, and C. Ambasana. 2010. Decolorization of water soluble azo dyes by bacterial cultures, isolated from dye house effluent. *Bioresource Technology* 101: 6580–6583.

Mohan, M. S., R. N. Chandrasekhar, K. K. Prasad, and J. Karthikeyan. 2002. Treatment of simulated Reactive Yellow 22 (azo) dye effluents using *Spirogyra* species. *Waste Management* 22: 575–582.

Moreira, S., A. M. F. Milagres, and S. I. Mussatto. 2014. Reactive dyes and textile effluent decolorization by a mediator system of salt-tolerant laccase from *Peniophora cinerea*. *Separation and Purification Technology* 135: 183–189.

Morrison, J. M., C. M. Wright, and G. H. John. 2012. Identification, isolation and characterization of a novel azoreductase from *Clostridium perfringens*. *Anaerobe* 18: 229–234.

Moya, R., M. Hernández, A. B. García-Martín, A. S. Ball, and M. E. Arias. 2010. Contributions to a better comprehension of redox-mediated decoloration and detoxification of azo dyes by a laccase produced by *Streptomyces cyaneus* CECT 3335. *Bioresource Technology* 3 (7): 2224–2229.

Mrozik, A., and S. Z. Piotrowska. 2010. Bioaugmentation as a strategy for cleaning up of soils contaminated with aromatic compounds. *Microbiology Research* 165 (5): 363–375.

Nadagouda, M. N., A. B. Castle, R. C. Murdock, S. M. Hussain, and R. S. Varma. 2010. *In vitro* biocompatibility of nanoscale zerovalent iron particles (NZVI) synthesized using tea polyphenols. *Green Chemistry* 12: 114–122.

Nakanishi, M., C. Yatome, N. Ishida, and Y. Kitade. 2001. Putative ACP phosphodiesterase gene (acpD) encodes an azoreductase. *Journal of Biological Chemistry* 276: 46394–46399.

Namdhari, B. S., S. K. Rohilla, R. K. Salar, S. K. Gahlawat, P. Bansal, and A. K. Saran. 2012. Decolorization of Reactive Blue MR, using *Aspergillus species* isolated from textile waste water. *ISCA Journal of Biological Sciences* 1 (1): 24–29.

Narayanan, K. B., and H. H. Park. 2015. Homogeneous catalytic activity of gold nanoparticles synthesized using turnip (*Brassica rapa* L.) leaf extract in the reductive degradation of cationic azo dye. *Korean Journal of Chemical Engineering* 32 (7): 1273–1277.

Nilsson, I., A. Moller, B. Mattiasson, M. Rubindamayugi, and U. Welander. 2006. Decolorization of synthetic and real textile wastewater by the use of white-rot fungi. *Enzyme and Microbial Technology* 38 (1–2): 94–100. doi: https://doi.org/10.1016/j.enzmictec.2005.04.020.

Nishiya, Y., and Y. Yamamoto. 2007. Characterization of a NADH: dichloroindophenol oxidoreductase from *Bacillus subtilis*. *Bioscience Biotechnology Biochemistry* 71: 611–614.

Njagi, E. C., H. Huang, L. Stafford, Genuino H, H. M. Galindo, J. B. Collins, G. E. Hoag et al. 2011. Biosynthesis of iron and silver nanoparticles at room temperature using aqueous sorghum bran extracts. *Langmuir* 27: 264–271.

Ogugbue, C. J., and T. Sawidis. 2011. Optimization of process parameters for bioreduction of azo dyes using *Bacillus firmus* under batch anaerobic condition. *International Journal of Environmental Studies* 68: 651–665.

Ola, I. O., A. K. Akintokun, I. Akpan, I. O. Omomowo, and V. O. Areo. 2010. Aerobic decolourization of two reactive azo dyes under varying carbon and nitrogen source by *Bacillus cereus*. *African Journal of Biotechnology* 9: 672–677.

Ong, S. A., K. Udhiyama, D. Inadama, Y. Ishida, and K. Yamagiwa. 2009. Phytoremediation of industrial effluent containing azo dye by model up flow constructed wetland. *Chinese Chemical Letters* 20: 225–228.

Ooi, T., T. Shibata, R. Sato, H. Ohno, S. Kinoshita, T. L. Thuoc, and S. Taguchi. 2007. An azoreductase, aerobic NADH-dependent flavo-protein discovered from *Bacillus* sp: Functional expression and enzymatic characterization. *Applied Microbiology and Biotechnology* 75: 377–386.

Ozer, A., G. Akkaya, and M. Turabik. 2006. The removal of Acid Red 274 from wastewater combined: Biosorption and bioaccumulation with *Spirogyra rhizopus*. *Dyes and Pigments* 71: 83–89.

Pak, R. Z., and S. S. Ardakani. 2016. Evaluation of kinetic and equilibrium parameters of NiFe2O4 NPs on adsorption of Reactive Orange dye from water. *Iranian Journal of Toxicology* 10 (2): 51–58.

Pal, S., U. Sarkar, and D. Dasgupta. 2010. Dynamic simulation of sec-ondary treatment processes using trickling filters in a sewage treatment works in Howrah, West Bengal, India. *Desalination* 253 (1): 135–140.

Pandey, A., P. Singh, and L. Iyengar. 2007. Bacterial decolorization and degradation of azo dyes. *International Biodeterioration & Biodegradation* 59: 73–84. doi: https://doi.org/10.1016/j.ibiod. 2006.08.006.

Pant, D., A. Singh, B. G. Van, O. S. Irving, N. P. Singh, L. Diels, and K. Vanbroekhoven. 2012. Bioelectrochemical systems (BES) for sus-tainable energy production and product recovery from organic wastes and industrial wastewaters. *RSC Advances* 2: 1248–1263.

Paszczynski, A., and R. C. Crawford. 1995. Potential for bioremediation of xenobiotic compounds by the white-rot fungus *Phanerochaete chrysosporium*. *Biotechnology Progress* 11: 368–379.

Patil, P., N. Desai, S. Govindwar, J. P. Jadhav, and V. Bapat. 2009. Degradation analysis of Reactive Red 198 by hairy roots of *Tagetes patula* L. (Marigold). *Planta* 230: 725–735.

Patrick, W., and S. Dietmar. 2007. Photodegradation of Rhodamine B in aqueous solution via SiO2 @ TiO2 nano-spheres. *Journal of Phytochemistry and Photobiology A: Chemistry* 185: 19–25.

Pelegrini, R., P. Peralto-Zamora, A. R. De-Andrade, J. Ryers, and N. Duran. 1999. Electrochemically assisted photocatalytic degradation of reactive dyes. *Applied Catalysis B: Environment* 22: 83–90.

Pokharia A., and S. S. Ahluwalia. 2016a. Biodecolorization and deg-radation of xenobiotic azo dye Basic Red 46 by *Staphylococcus epidermidis* MTCC 10623. *International Journal of Research in Biosciences* 5 (2): 10–23.

Pokharia, A., and S. S. Ahluwalia. 2016b. Decolorization of xenobiotic azo dye- Black WNN by immobilized *Paenibacillus alvei* MTCC

10625. *International Journal of Environmental Bioremediation & Biodegradation* 4 (2): 35–46.

Pouretedal, H. R., and M. H. Keshavarz. 2011. Study of Congo Red photodegradation kinetic catalyzed by Zn1-xCuxS and Zn1-x S nanoparticles. *International Journal of the Physical Sciences* 6 (27): 6268–6279.

Prasad, A. S. A., and K. V. B. Rao. 2013. Aerobic biodegradation of azo dye by *Bacillus cohnii* MTCC 3616; an obligately alkaliphilic bacterium and toxicity evaluation of metabolites by different bioassay systems. *Applied Microbiology and Biotechnology* 97: 7469–7481.

Praveen, G. N. K., and S. K. Bhat. 2012. Fungal degradation of azo dye Red 3BN and optimization of physico-chemical parameters. *ISCA Journal of Biological Sciences* 1 (2): 17–24.

Przystas, W., E. Zabłocka-Godlewska, and E. Grabinska-Sota. 2018. Efficiency of decolorization of different dyes using fungal biomass immobilized on different solid supports. *Brazilian Journal of Microbiology* 49: 285–295.

Qiu, B., Y. Dang, X. Cheng, and D. Sun. 2013. Decolorization and degradation of cationic red X-GRL by upflow blanket filter. *Water Science and Technology* 67 (5): 976–982.

Rabaey, K., J. Rodriguez, L. L. Blackall, J. Keller, P. Gross, D. Batstone, W. Verstraete, and K. H. Nealson. 2007. Microbial ecology meets electrochemistry: Electricity driven and driving communities. *ISME Journal* 1: 9–18.

Ramalho, P. A., S. Paiva, A. Cavaco-Paulo, M. Casal, M. H. Cardoso and M. T. Ramalho. 2005. Azoreductase activity of intact *Saccharomyces cerevisiae* cells is dependent on the Fre1p component of plasma membrane ferric reductase. *Applied and Environmental Microbiology* 71: 3882–3888.

Rammel, R. S., S. A. Zatiti, and E. M. M. Jamal. 2011. Biosorption of Crystal Violet by *Chaetophora elegans* alga. *Journal of the University of Chemical Technology and Metallurgy* 46 (3): 283–292.

Ranganathan, K., K. Karunagaran, and D. C. Sharma. 2007. Recycling of wastewaters of textile dyeing industries using advanced treatment technology and cost analysis—Case studies. *Resources Conservation and Recycling* 50: 306–318.

Rieger, P. G., H. M. Meier, M. Gerle, U. Vogt, T. Groth, and H. J. Knackmuss. 2002. Xenobiotics in the environment: Present and future strategies to obviate the problem of biological persistence. *Journal of Biotechnology* 94: 101–23.

Russ, R., J. Rau, and A. Stolz. 2000. The function of cytoplasmic flavin reductases in the reduction of azo dyes by bacteria. *Applied and Environmental Microbiology* 66: 1429–1434.

Ryan, A., N. Laurieri, I. Westwood, C. J. Wang, E. Lowe, and E. Sim. 2010. A novel mechanism for azoreduction. *Journal of Molecular Biology* 400: 24–37.

Safavi, A., and S. Momeni. 2012. Highly efficient degradation of azo dyes by palladium/hydroxyapatite/Fe3O4 nanocatalyst. *Journal of Hazardous Materials* 30 (201–202): 125–31.

Sandhya, S., K. Sarayu, B. Uma, and K. Swaminathan. 2008. Decolorizing kinetics of a recombinant *Escherichia coli* SS125 strain harboring azoreductase gene from *Bacillus latrosporus* RRK1. *Bioresource Technology* 99 (7): 2187–2191.

Saratale, R. G., G. D. Saratale, D. C. Kalyani, J. S. Chang, and S. P. Govindwar. 2009. Enhanced decolorization and biodegradation of textile azo dye Scarlet R by using developed microbial consortium-GR. *Bioresource Technology* 100: 2493–2500.

Saratale, R. G., G. D. Saratale, J. S. Chang and S. P. Govindwar. 2011. Bacterial decolorization and degradation of azo dyes: A review. *Journal of the Taiwan Institute of Chemical Engineers* 42: 138–157.

Sarayu, K., and S. Sandhya. 2012. Current technologies for biological treatment of textile wastewater–A review. *Applied Biochemistry and Biotechnology* 167 (3): 645–661. doi: 10.1007/s12010-012-9716-6.

Satiroglu, N., Y. Yalcinkaya, A. Denizli, M. Y. Arica, S. Bektas, and O. Genc. 2002. Application of NaOH treated *Polyporus versicolor* for removal of divalent ions of group IIB elements from synthetic wastewater. *Process Biochemistry* 38: 65–72.

Sayler, G. S., and S. Ripp 2000. Field applications of genetically engineered microorganisms for bioremediation processes. *Current Opinion in Biotechnology* 11: 286–289.

Selvam, K., and P. M. Shanmuga. 2012. Biological treatment of azo dyes and textile industry effluent by newly isolated white rot fungi *Schizophyllum commune* and *Lenzites eximia*. *International Journal of Environmental Sciences* 2 (4): 1926–1935.

Selvam, K., K. Swaminathan, and C. Keo-Sang. 2003. Microbial decolorization of azo dyes and dye industry effluent by *Fomes lividus*. *World Journal of Microbiology and Biotechnology.* 19: 591–593.

Shabbir, S., M. Faheem, N. Ali, P. G. Kerr, and Y. Wu. 2017. Periphyton biofilms: A novel and natural biological system for the effective removal of sulphonated azo dye Methyl Orange by synergistic mechanism. *Chemosphere* 167: 236–246.

Sharma, K. P., K. Sharma, S. Kumar, S. Sharma, R. Grover, P. Soni, and S. M. Bharadwaj et al. 2005. Response of selected aquatic macrophytes towards textile dye wastewaters. *Indian Journal of Biotechnology* 4: 538–545.

Sharma, P., R. Goel, and N. Caplash. 2007. Bacterial laccases. *World Journal of Microbiology and Biotechnology* 23: 823–832.

Shinde, U. G, S. K. Metkar, R. L. Bodkhe, G. Y. Khosare, and S. N. Harke. 2012. Potential of polyphenol oxidases of *Parthenium hysterophorus, Alternanthera sessilis* and *Jotrapha curcas* for simultaneous degradation of two textiles dyes: Yellow 5G and Brown R. *Trends in Biotechnology Research* 1: 24–28.

Singh, P. K., and R. L. Singh. 2017. Bio-removal of azo dyes: A review. *International Journal of Applied Sciences and Biotechnology* 5 (2): 108–126.

Singh, R. L. 1989. Metabolic disposition of (14-C) Metanill Yellow in rats. *Biochemistry International* 19: 1109–1116.

Singh, R. L., R. Gupta, and R. P. Singh. 2015a. Microbial degradation of textile dyes for environmental safety. In *Advances in Biodegradation and Bioremediation of Industrial Waste*, edited by R. Chandra, 249–285. CRC Press.

Singh, R. L., P. K. Singh, and R. P. Singh. 2015b. Enzymatic decolorization and degradation of azo dyes-a review. *International Biodeterioration & Biodegradation* 104: 21–31.

Singh, R. L., S. K. Khanna, and G. B. Singh. 1987. Safety evaluation studies on pure Metanil Yellow acute and sub-chronic exposure responses. *Beverage and Food World* 74: 9–13.

Singh, R. L., S. K. Khanna, and G. B. Singh. 1988. Acute and short term toxicity of popular blend of Metanil Yellow and Orange II in albino rats. *Indian Journal of Experimental Biology* 26: 105–111.

Singh, R. L., S. K. Khanna, and G. B. Singh. 1991. Metabolic disposition of (14-C) Metanil Yellow in guinea pigs. *Veterinary and Human Toxicology* 33 (3): 220–223.

Singh, R. P., P. K. Singh, and R. L. Singh. 2014. Bacterial decolorization of textile azo dye Acid Orange by *Staphylococcus hominis* RMLRT03. *Toxicology International* 21(2): 160–166.

Singh, R. P., P. K. Singh, and R. L. Singh. 2017a. Present status of biodegradation of textile dyes. *Current Trends in Biomedical Engineering & Biosciences* 3(4): 555618. doi:10.19080/CTBEB.2017.03. 555618.

Singh, R. P., P. K. Singh, and R. L. Singh. 2017b. Role of azoreductases in bacterial decolorization of azo dyes. *Current Trends in*

Biomedical Engineering & Biosciences 9 (3): 555764. doi:10.19080/CTBEB.2017.09.555764.

Singh, R. P., P. K. Singh, R. Gupta, and R. L. Singh. 2019. Treatment and recycling of wastewater from textile industry. In Advances in Biological Treatment of Industrial Waste Water and their Recycling for a Sustainable Future edited by R. L. Singh and R. P. Singh, 225–266. Springer Nature.

Solis, M., A. Solis, H. I. Perez, N. Manjarrez, and M. Flores. 2012. Microbial decoloration of azo dyes: A review. *Process Biochemistry* 47: 1723–1748. doi: http://dx.doi.org/10.1016/j.procbio.2012.08.014.

Stolz, A. 2001. Basic and applied aspects in the microbial degradation of azo dyes. *Applied Microbiology and Biotechnology* 56: 69–80.

Suganya, S., and K. Revathi. 2016. Decolorization of reactive dyes by immobilized bacterial cells from textile effluents. *International Journal of Curruent Microbiology and Applied Sciences* 5 (1): 528–532.

Sugumar, S., and B. Thangam. 2012. BiodEnz: A database of biodegrading enzymes. *Bioinformation* 8: 40–42.

Sultana, S., S. Ramabadran, and P. R. Swathi. 2015. Photocatalytic degradation of azo dye using ferric oxide nanoparticles. *International Journal of Chem Tech Research* 8 (3): 1243–1247.

Sun, J., Y. Li, Y. Hu, B. Hou, Y. Zhang, and S. Li. 2013. Understanding the degradation of Congo Red and bacterial diversity in an air cathode microbial fuel cell being evaluated for simultaneous azo dye removal from wastewater and bioelectricity generation. *Applied Microbiology and Biotechnology* 97: 3711–3719.

Sun, J., Y. Y. Hua, Z. Bi, and Y. Q. Cao. 2009. Simultaneous decolorization of azo dye and bioelectricity generation using a microfiltration membrane air cathode single chamber microbial fuel cell. *Bioresource Technology* 100: 3185–3192.

Sun, J., Z. Bi, B. Hou, and Y. Cao. 2011. Further treatment of decolorization liquid of azo dye coupled with increased power production using microbial fuel cell equipped with an aerobic biocathode. *Water Research* 45 (1): 283–291.

Sunkar, S., and C. Vallinachiyar. 2013. Endophytic fungi mediated extracellular silver NPs as effective antibacterial agents. *International Journal of Pharmacy and Pharmaceutical Sciences* 5 (2): 95–100.

Suzuki, Y., T. Yoda, A. Ruhul, and W. Sugiura. 2001. Molecular cloning and characterization of the gene coding for azoreductase from *Bacillus* sp. OY1-1 isolated from soil. *Journal of Biological Chemistry* 276: 9059–9065.

Tadjarodi, A., M. Imani, M. Kerdari, K. Bijanzad, D. Khaledi, and M. Rad. 2014. Preparation of CdO Rhombus-like nanostructure and its photocatalytic degradation of azo dye from aqueous solution. *Nanomaterials and Nanotechnology* 4 (1): 1. doi: 10.5772/58464.

Tahir, U., S. Sohail, and U. H. Khan. 2017. Concurrent uptake and metabolism of dyestuffs through bio-assisted phytoremediation: A symbiotic approach. *Environment Science and Pollution Research International* 29: 22914–22931.

Telke, A. A., G. S. Ghodake, D. C. Kalyani, R. S. Dhanve, and S. P. Govindwar. 2011. Biochemical characteristics of a textile dye degrading extracellular laccase from a *Bacillus sp.* ADR. *Bioresource Technology* 102: 1752–1756.

Telke, A. A., S. M. Joshi, S. U. Jadhav, D. P. Tamboli, and S. P. Govindwar. 2010. Decolorization and detoxification of Congo Red and textile industry effluent by an isolated bacterium *Pseudomonas* sp. SU-EBT. *Biodegradation* 21: 283–296.

Theerachat, M., S. Morel, D. Guieysse, M. R. Simeon, and W. Chulalaksananukul. 2012. Comparison of synthetic dye decolorization by whole cells and a laccase enriched extract from *Trametes versicolor* DSM11269. *African Journal of Biotechnology* 11 (8): 1964–1969.

Van der Zee, F. P. 2002. Anaerobic azo dye reduction. Doctoral Thesis, Wageningen University, Wageningen, The Netherlands, p.142.

Vandevivere, P. C., R. Bianchi, and W. Verstraete. 1998. Treatment and reuse of wastewater from the textile wet-processing industry: Review of emerging technologies. *Journal of Chemical Technology and Biotechnology* 72: 289–302.

Vijayakumar, S., and C. Manoharan. 2012. Treatment of dye industry effluent using free and immobilized cyanobacteria. *Journal of Bioremediation and Biodegradation* 3: 10. doi: 10.4172/2155-6199.1000165.

Wang, C. J., C. Hagemeier, N. Rahman, E. Lowe, M. Noble, M. Coughtrie, and E. Sim. 2007. Westwood I: Molecular cloning, characterisation and ligand bound structure of an azoreductase from *Pseudomonas aeruginosa*. *Journal of Molecular Biology* 373 (5): 1213–1228.

Wang, W., Z. Zhen, N. Hong, Y. Xiaomeng, L. Qianqian, and L. Lin. 2012. Decolorization of industrial synthetic dyes using engineered *Pseudomonas putida* cells with surface immobilized bacterial laccase. *Microbial Cell Factories* 11: 75. doi: 10.1186/1475-2859-11-75.

Wang, Y. Z., A. Zhou, Y. H. Xie, J. Han, M. Y. You, M. Wang, and T. Zhu. 2016. Azo dye decolorization in bioelectrochemical system: Characteristic analysis of electrochemical active biofilms. *International Journal of Electrochemical Sciences* 11: 7947–7959.

Wang, Z. W., J. S. Liang, and Y. Liang. 2013. Decolorization of Reactive Black 5 by a newly isolated bacterium *Bacillus* sp. YZU1. *International Biodeterioration & Biodegradation* 76: 41–48.

Watanabe, K., M. Manefield, M. Lee, and A. Kouzuma. 2009. Electron shuttles in biotechnology. *Current Opinion in Biotechnology* 20 (6): 633–641.

Xu, L., and J. Wang. 2011. A heterogeneous Fenton-like system with nanoparticulate zerovalent iron for removal of 4-chloro-3-methyl phenol. *Journal of Hazardous Materials* 186: 256–264.

Xu, L., X. C. Wu, and J. J. Zhu. 2008. Green preparation and catalytic application of Pd nanoparticles. *Nanotechnology* 19 (30): 305603. doi.org/10.1088/0957-4484/19/30/305603.

Xu, Y., and R. E. Leburn. 1999. Treatment of textile dye plant effluent by nanofiltration membrane. *Separation Science and Technology* 34: 2501–2519.

Yang, G., Y. Liu, and Q. Kong. 2004. Effect of environmental factors on dye decolorization by *P. sordida* ATCC 90872 in an aerated reactor. *Process Biochemistry* 39: 1401–1405.

Yeh, M. S., and J. S. Chang. 2004. Bacterial decolorization of an azo dye with a natural isolates of *Pseudomonas luteola* and genetically modified *Escherichia coli*. *Journal of Chemical Technology & Biotechnology* 79: 1354–1360. doi:10.1002/jctb.1099.

Yogendra, K., K. M. Mahadevan, N. Madhusudhana, S. Naik and H. Gopalappa. 2011. Photocatalytic decolorization of Coralene Dark Red 2B azo dye by using calcium zincate NPs in the presence of natural sunlight. International Conference on Chemistry and Chemical Process IACSIT Press, Singapore 10: 8–12.

Zhang, B., and Y. Zhu. 2011. Simultaneous decolorization and degradation of azo dye with electricity generation in microbial fuel cells. *Second International Conference on Mechanic Automation and Control Engineering*. doi: 10.1109/MACE.2011.5987508.

Zhang, X., Y. Liu, K. Yan, and H. Wu. 2007. Decolorization of anthraquinone-type dye by bilirubin oxidase-producing non-ligninolytic fungus *Myrothecium* sp. IMER1. *Journal of Bioscience and Bioengineering* 104 (2): 104–110.

Zheng, M., Y. Chi, H. Yi, and S. Shao. 2014. Decolorization of Alizarin Red and other synthetic dyes by a recombinant laccase from *Pichia pastoris*. *Biotechnology Letters* 36: 39–45.

Index

For Product Safety Concerns and Information please contact our EU
representative GPSR@taylorandfrancis.com Taylor & Francis Verlag GmbH,
Kaufingerstraße 24, 80331 München, Germany

Printed and bound by CPI Group (UK) Ltd, Croydon, CR0 4YY
08/05/2025
01864439-0002